Eastern Snow Conference 2018

College Park, Maryland, USA
5 - 8 June 2018

ISBN: 978-1-7138-0355-3

Printed from e-media with permission by:

Curran Associates, Inc.
57 Morehouse Lane
Red Hook, NY 12571

Some format issues inherent in the e-media version may also appear in this print version.

Eastern Snow Conference
C/O Dr. Krystopher Chutko
117 Science Pl-Dept. Geography
Saskatoon, Sk, Canada S7N 5C8

https://www.easternsnow.org/

Additional copies of this publication are available from:

Curran Associates, Inc.
57 Morehouse Lane
Red Hook, NY 12571 USA
Phone: 845-758-0400
Fax: 845-758-2633
Email: curran@proceedings.com
Web: www.proceedings.com

Proceedings of the
75th ANNUAL

EASTERN SNOW CONFERENCE

5-8 June 2018

NOAA Center for Weather and Climate Prediction
College Park, Maryland, USA

Dr. Peter Adams
1936-2018

The Proceedings of the 75th Eastern Snow Conference is dedicated to Life Member Peter Adams, who passed away on September 28, 2018 in Peterborough, Ontario, Canada, at the age of 82.

Dr. Adams' achievements include more than 60 peer-reviewed articles and several books, mostly related to snow and ice research and in particular to his glaciological work on Axel Heiberg Island, which is the longest continuous study of its kind in the High Arctic. Born in England, he moved to Montreal in 1959 with a Carnegie Arctic Scholarship, and completed his PhD in Geography and Glaciology at McGill University. In 1968, he became the founding Chair of the Department of Geography at Trent University. Later in life, his attention turned to politics at both the provincial and federal levels, promoting and advancing science and scholarship. After retirement, Peter and his wife Jill devoted their time to volunteer work with children around the world.

Peter was ESC President in 1984-85 and won the Sno-Foo Award in 1986 for his impromptu "yellow snow" experiment during a snow sculpture event on a U.S. university campus.

He will be greatly missed.

CONTENTS

Session #3: Snow Modeling & Snow Processes

Poster Session

Session #5: Snow Water Equivalent and Watershed Hydrology

Session #6: Snow and Ice Remote Sensing

FOREWORD

This proceedings volume contains papers presented at the 75th Eastern Snow Conference (ESC) held 5-8 June 2018 in College Park, Maryland, USA. The meeting featured oral and poster sessions on recent advances in remote sensing, remote sensing theory, applications of remote sensing ranging from the ice sheets to the tropics, snow and ice processes, and hydroclimatology.

The ESC is a joint United States and Canadian forum for discussing recent work on operational, applied and scientific issues related to snow and ice. It also retains an increasing interest as a symposium where novel approaches to cryospheric science of international significance are presented. The ESC has published an annual series of proceedings since 1952. Typical topics include studies of snow and ice as materials, snow removal, meteorological forecasting, river ice control, snow hydrology, snow chemistry, glaciology, remote sensing of snow and ice, and snow ecology. Membership in the ESC is open to all interested individuals and corporations. Additional copies of the current proceedings and all back issues can be obtained from the Secretary. Additional information about the Eastern Snow Conference may be found at: http://www.easternsnow.org/.

We thank members of the ESC executive committee (especially Mauri Pelto) for outstanding effort in publishing ESC 2017 proceedings in *Physical Geography*. Over the years, the proceedings committee has contributed considerable time to this publication and has enhanced the quality of these proceedings.

The 2018 meeting of the Eastern Snow Conference and these Proceedings were made possible in part by Corporate membership from Campbell Scientific Canada, Geonor, and Hoskin Scientific Ltd. The National Oceanic and Atmospheric Administration's Center for Weather and Climate Prediction provided meeting space, technical, and logistic support.

Campbell Scientific (Canada) Corp.
Edmonton, Alberta, Canada
http://www.campbellsci.ca

Hoskin Scientific Limited
Burnaby, British Columbia, Canada
http://www.hoskin.ca

GEONOR
Augusta, New Jersey, USA
http://www.geonor.com

Center for Weather and Climate Prediction
National Oceanic and Atmospheric Administration
College Park, MD, USA

We look forward to seeing you at the 76th Eastern Snow Conference meeting to be held 4–6 June 2019 at the Lake Morey Resort in Fairlee, Vermont, USA.

Krystopher Chutko and George Riggs
ESC Proceedings Co-Editors
University of Saskatchewan and NASA GSFC/SSAI

STATEMENT OF PURPOSE

The Eastern Snow Conference (ESC) is a joint Canadian/US organization founded in the 1940s, originally with members primarily from eastern North America. Our current members are scientists, snow surveyors, engineers, technicians, professors, students, and operational and maintenance professionals from North America, the United Kingdom, Japan, and Germany. There is a western counterpart to the ESC, the Western Snow Conference (WSC), which also is a joint Canadian/US organization. Every fifth year, the ESC and the WSC hold joint meetings.

The Eastern Snow Conference is a forum that brings the research and operations communities together to discuss recent work on scientific, applied, and operational issues related to snow and ice. The location of the conference alternates yearly between the United States and Canada, and attendees present their work by either giving a talk or presenting a poster. Most resulting papers are reviewed, edited, and published in our yearly *Proceedings of the Eastern Snow Conference.* In recent years, the ESC meetings have included sessions on snow physics, winter survival of animals, snow and ice loads on structures, river ice, remote sensing of snow and ice, and glacier processes. Volumes of the *Proceedings* can be found in libraries throughout North America and Europe, and the papers are also available through the National Technical Information Service (NTIS) in the United States and CISTI in Canada.

* * * * * * * * *

Le Colloque sur la neige région Est (ESC) est une organisation américaine-canadienne fondée dans les années '40 et dont les membres provenaient à l'origine principalement de l'est de l'Amérique-du-Nord. Actuellement, les membres, qu'ils soient chercheurs, techniciens, ingénieurs, professeurs, étudiants, et spécialistes des services d'exploitation et d'entretien, viennent non seulement d'Amérique-du-Nord, mais aussi du Royaume-Uni, du Japon, et d'Allemagne. Le Colloque sur la neige-région Ouest (WSC), aussi une organisation américaine-canadienne, est l'homologue de l'ESC pour l'ouest nord-américain. Tous les cinq ans, l'ESC et la WSC organisent des réunions en commun.

L'ESC est un forum qui rassemble chercheurs et responsables des services d'exploitation pour discuter des travaux récents sur les problèmes scientifiques, opérationnels, ou autres dus à la neige et à la glace. Le site de cette réunion annuelle alterne entre les États Unis et le Canada. Les participants y présentent les résultats de leurs travaux par des communications orales ou au moyen d'affiches. Ces communications, une fois revues et éditées, sont publiées dans les *Annales de l'ESC.* Dans les années récentes, les réunions de l'ESC ont inclus des sessions sur la physique de la neige, la survie hivernale de la faune, les forces exercées par la neige et la glace sur les structures et les bâtiments, la glace de rivière, la télédétection de la neige et de la glace, et les processus glaciaires. Les *Annales de l'ESC* sont accessibles dans la plupart des bibliothèques scientifiques d'Amérique-du-Nord et d'Europe. Des copies d'articles peuvent être obtenues du National Technical Information Service (NTIS) aux États Unis et son équivalent au Canada, le CISTI.

EXECUTIVES FOR THE 2017-18 75th EASTERN SNOW CONFERENCE

PRESIDENT

Alexandre Roy
Sherbrooke
Quebec, Canada

PAST PRESIDENT

Michael Durand
Columbus
Ohio, USA

VICE-PRESIDENT AND PROGRAM CHAIR

George Riggs
Gambrills
Maryland, USA

SECRETARY-TREASURER (Canada)

Miles Ecclestone
Peterborough
Ontario, Canada

Krystopher Chutko
Saskatoon
Saskatchewan, Canada

Alexandre Langlois
Sherbrooke
Quebec, Canada

SECRETARY-TREASURER (USA)

Ken Rancourt
Conway
New Hampshire, USA

Derrill Cowing
Monmouth
Maine, USA

ESC PROCEEDINGS EDITORS

Alexandre Langlois
Sherbrooke
Quebec, Canada

Krystopher Chutko
Saskatoon
Saskatchewan, Canada

***PHYSICAL GEOGRAPHY* EDITORS**

Mauri Pelto (Chair)
Dudley
Massachusetts, USA

Robert Hellström
Bridgewater
Massachusetts, USA

STEERING COMMITTEE

Eli Deeb
Hanover
New Hampshire, USA

Chris Fuhrman
Chapel Hill
North Carolina, USA

Steve Howell
Toronto
Ontario, Canada

Craig Smith
Saskatoon
Saskatchewan, Canada

Laura Thomson
Ottawa
Ontario, Canada

Carrie Vuyovich
Hanover
New Hampshire, USA

RESEARCH COMMITTEE

Sean Helfrich (Chair)
Suitland
Maryland, USA

James Brylawski
Augusta
New Jersey, USA

Barton Forman
College Park
Maryland, USA

David Robinson
New Brunswick
New Jersey, USA

Joan Ramage
Bethlehem
Pennsylvania, USA

Kevin Kang
Waterloo
Ontario, Canada

WEBMASTER

Vincent Sasseville
Sherbrooke
Quebec, Canada

LOCAL ARRANGEMENTS

Sean Helfrich
College Park
Maryland, USA

Barton Forman
College Park
Maryland, USA

George Riggs
Gambrills
Maryland, USA

THE PRESIDENT'S PAGE

The 75th annual meeting of the Eastern Snow Conference was held at the NOAA Center for Weather and Climate Prediction, Climate Prediction Center, College Park, Maryland, USA from June 5 - 8 2018. The NOAA Center for Weather and Climate Prediction in College, houses dedicated scientists who provide the United States with expert weather, water, and climate forecasts that touch everyone's lives. Opened in August 2012, the center provides a seamless suite of environmental analysis, diagnostics and forecasts from the surface of the sun to the depths of the ocean floor. This integrated facility advances NOAA's effort to build a Weather-Ready Nation through improved precision forecasts, effective communications and strong partnerships.

The scientific program included 34 oral presentations and 29 poster presentations. The theme of this year's program was "*Snow past present and future*". Once again, the student participation was excellent, representing nearly 40% of the attendees and resulting in very stimulating presentations. The Research Committee at the Banquet recognized the following students for exemplary achievement:

- Jiyue Zhu from the University of Michigan received the Wiesnet Medal for the best student paper entitled "Retrieval algorithm of snow water equivalent using SnowSAR and scatterometer backscatters with both co- and cross-polarizations".
- Justin M. Pflug from the University of Washington received the David Hewitt Miller Best Student Poster Award for his team's poster entitled "Adapting model representation of liquid water percolation in maritime environments".

We were very fortunate to have Dr. Louis Uccellini, NOAA Assistant Administrator for Weather Services and Director National Weather Service, deliver the Banquet Address. His address presented a very interesting overview of snowstorm forecasting in the last decades.

Ronny Schroeder presented the coveted Sno-Foo Award to Alexandre Langlois from Université de Sherbrooke, for destroying a pickup truck door on a snow bank.

Dorothy Hall also introduced our new prestigious Life Member, Dr. James Foster.

A special thanks to George Riggs who was the Vice-President and Program Chair. Thanks to Sean Helfrich and Barton Forman for local arrangements. ESC Executive Committee Members Krys Chutko, Miles Ecclestone, Derrill Cowing and Ken Rancourt continue to provide outstanding leadership and were instrumental to the success of the 75th meeting. The efforts of webmaster Vincent Sasseville, and members of the steering and research committees are also greatly appreciated. Likewise, the meeting would not have been a success without the participation and continued support of our corporate members: Campbell Scientific Canada, Geonor, and Hoskin Scientific.

Please plan to attend the 76th Eastern Snow Conference at Lake Morey, Vermont, USA, from June 4 - 6 2019. Elias Deeb, Cold Regions Research and Engineering Laboratory, is the Vice-President and Program Chair. Students should check the ESC website for more information on submitting papers for the Wiesnet Medal.

Thank you for your participation and support of the Eastern Snow Conference and I look forward to seeing you all next year in Vermont. In the meantime, best wishes for a snowy, safe, and successful field season!

Alexandre Roy
75th President
Eastern Snow Conference

Dr. James Foster – ESC Life Member

The Eastern Snow Conference gratefully recognizes individuals who have made lifelong contributions to the study of snow and ice, and for their support of this organization. At the 75th Eastern Snow Conference, Dr. James Foster was inducted as the newest Life Member, joining Peter Adams, Barry Goodison, Gerry Jones, John Metcalfe, Hilda Snelling, and Donald Wiesnet.

Induction presented by Dorothy Hall

It's my pleasure to introduce Jim Foster as a new Life Member of the Eastern Snow Conference!

Jim has made substantial and long-lasting contributions to snow and ice research. He has also worked to further the public understanding of Earth Science and is an enthusiastic member of the Eastern Snow Conference.

As Jim mentioned yesterday in his talk, his first Eastern Snow Conference was in 1978, at the 35th. Here's a picture of Jim that was taken around 1978. It's hard to believe that that was 40 years ago, because the 35th was my first ESC meeting, too! Jim remembers that Don Wiesnet, also a Life Member, encouraged us to attend that meeting which was held in early February at CRREL in Hanover, New Hampshire. With his remarkable memory of details, Jim also recalls the weather being very cold, but not snowy, during that meeting. In fact, that was the last ESC that was held during the winter. Since then they've all been held in the spring to avoid unpleasant distractions… like snowstorms.

At that 1978 meeting, Jim gave a presentation about the extensive ice cover on the Chesapeake Bay during the very cold winter of 1976-77. Along with two co-authors, Dorothy Schultz and Bill Dallam, he inferred ice-type and -thickness categories from Landsat images. Forty years later, Jim continues to study ice conditions in the Bay, as we heard in his very interesting talk yesterday just before the cruise.

During Jim's long and successful career with NASA / Goddard Space Flight Center, he focused on studying snow using satellites – mainly using Landsat visible/near-IR data in the early years, and later, passive-microwave data. He wrote a ground-breaking paper in 1989 noting the increasingly-earlier disappearance of snow cover in Arctic tundra areas. A huge body of literature on related topics has followed. He's also made substantial contributions to the difficult problem of assessing snow-water equivalent from space. Also, notably, Jim contributed to the study of snow crystals using scanning electron microscopy.

Jim has done field work in many cold places all over the world including Antarctica and Greenland, and in addition, he's been and continues to be very active in science outreach.

Although Jim never botched anything up badly enough to be a recipient of the coveted Sno-Foo award, over the last 40 years he's been a big supporter and frequent presenter at the ESC. Jim served as President in 1989, and presided over the 46th annual meeting, which was held in Quebec City.

On behalf of the current Executive Committee I hereby induct you, Jim Foster, as a Life Member of the Eastern Snow Conference!

Wiesnet Medal

The Eastern Snow Conference encourages student research through its Wiesnet Medal. This medal is presented annually to the best student paper presented at the conference.

Year	Winner	Affiliation
2018	Jiyue Zhu	University of Michigan
2017	Caroline Dolant	Université de Sherbrooke
2016	Syed Mousavi	University of Michigan
2015	Nicolas Leroux	University of Saskatchewan
2014	Justin Hartnett	Syracuse University
2013	Andreas Dietz	Earth Observation Center/DFD, Germany
2012	Elizabeth Burakowski	University of New Hampshire
2011	Kathryn Semmens	Lehigh University
2010	Simon von de Wall	University of Victoria
2009	Si Chen	Dartmouth College
2008	Chris Fuhrmann	University of North Carolina at Chapel Hill
2007	*Not awarded*	
2006	Y.C. Chung	University of Michigan
2005	M. Javan-Mashmool	Université du Québec à Chicoutimi
2004	J. Farzaneh-Dehkordi	Université du Québec à Chicoutimi
2003	Alexandre Langlois	Université de Sherbrooke
2002	Patrick Ménard	Université de Laval
2001	C. Tavakoli	Université du Québec à Chicoutimi
2000	*Not awarded*	
1999	S. Brettschnieder	Université du Québec à Chicoutimi
1998	Andrew Grundstein	University of Delaware
1997	Newell Hedstrom	University of Saskatchewan
1996	Suzanne Hartley	University of Denver
1995	Paul Wolfe	Wilfred Laurier University
1994	G.E. Mann	University of Michigan
1993	G. Devarennes	Université du Québec à Québec
1992	D.W. Cline	University of Colorado
1991	D. Samelson	Cornell University
1990	A.K. Abdel-Zaher	University of New Brunswick
1989	A. Giguere	McGill University
1988	Mauri Pelto	University of Maine
1987	Cameron Wake	Wilfred Laurier University
1986	Craig Allan	Trent University
1985	Robert Speck	Rensselaer Polytechnic Institute
1984	N.K. Kalliomaki	Laurentian University
1983	David Beresford	Trent University
1982	*Not awarded*	
1981	Jeffrey Patch	University of New Brunswick
1980	Bryan Wolfe	Trent University
1979	Margaret Leech	McGill University
1978	Michael English	Trent University
1977	Don McLaughlin & George Duggan	Rensselaer Polytechnic Institute
1976	Dwayne McMurter	Trent University
1975	Nigel Allan	Syracuse University
1974	*Not awarded*	
1973	Stan Mathewson	Trent University

David Miller Award

The David Miller Award is presented to the best student poster at the annual Conference.

Year	Winner	Affiliation
2018	Justin Pflug	University of Washington
2017	Elizabeth Ryan	University of Maryland
2016	Oliver Wigmore	The Ohio State University
2015	Charles Papasodoro	Université de Sherbrooke
2014	Laura Thomson	University of Ottawa
2013	Nastaran Saberi	University of Waterloo
2012	Johnathan Sugg	Appalachian State University
2011	*Not awarded*	
2010	Nicolas Svacina	University of Waterloo
2009	Iliyana Dobreva	Texas A&M University
2008	Christopher Karmosky	Pennsylvania State University

Campbell Scientific Canada Prize

The Campbell Scientific Canada prize is awarded to student research showing the most innovative use of technology in the gathering of data.

Year	Winner	Affiliation
2018	*Not awarded*	
2017	Abigail Dalton	University of Ottawa
2016	*Not awarded*	
2015	*Not awarded*	
2014	Aaron Campbell	University of Waterloo
2013	*Not awarded*	
2012	Andrew Kasurak	University of Waterloo
2011	Mathieu Beaulieu	University of British Columbia
2010	*Not awarded*	
2009	*Not awarded*	
2008	Nicholas Kinar	University of Saskatchewan
2007	Gro Lilbaek	University of Saskatchewan
2006	Gernot Koboltschnig	University of Natural Resources and Applied Life Sciences, Vienna
2005	Warren Helgason	University of Saskatchewan

Session #1: Advances in Remote Sensing

Remote Sensing of Snow: A Four Act Play

RICHARD KELLY[1]

ABSTRACT

Remote sensing can be used effectively to address big science questions regarding the role of snow in the hydrological cycle. Remote sensing has helped us to understand how snow moderates the planetary energy budget through its reflective properties. It has also helped to provide estimates of increasingly critical water supply storage in snowmelt-dominated catchments. Remote sensing, therefore, can be used to characterize "more or less" where snow is and how much is there. In addition, the question about when has snow reached a ripening phase can also be considered using remote sensing observations.

But the "more or less" clause above is important and perhaps signifies qualified equivocation in the use of remote sensing for snow hydrologic science. The root of this equivocation is a function of the history of how remote sensing has evolved for estimating snow. In essence there has been no dedicated snow remote sensing mission. Snow scientists have been required to leverage instruments of opportunity for their means which has led to the "if you build it they will come" approach to remote sensing of snow. Yet progress can only be made so far; these instruments of opportunity were designed for completely different applications and the measurement physics were never tuned to snow. Hence the "more or less" clause. However, over the years, snow scientists have been keenly interested the interplay between field-based snow research and technological innovation which has led to advances in: (i) in situ field-based snow measurement technologies; (ii) snow remote sensing observation technologies; (iii) snow modelling technologies (physics of radiation interactions with snow and snow hydrology).

From the initial impetus of using instruments of opportunity to the active exploration of innovative techniques, this paper characterizes the development of remote sensing of snow in a manner that could be seen as remarkably similar to the structure of a four act play. Act 1: "the inciting incident". An existing satellite observing instrument, specifically designed and operated for another purpose, is deemed to hold promise for observing snow. Science is set along a path to explore remote sensing as a viable hydrological measurement tool. Act 2: "the pinch point". Progress is constricted at the realization that the nature of the remote sensing of snow is more challenging than initially thought as science begins to uncover the complex electromagnetic response of snow. Act 3: "a turn for the worse". Field experiments and model representations of snow indicate snow to be significantly more complex than originally conceived in the form of an elaborate time-varying three dimensional electromagnetic medium that can only be accurately characterized by tuned observing instruments and mature modelling technologies. Act 4: "the denouement". Based on research and technological innovation, a new mission is conceived that surpasses the "more or less" clause and provides high quality and robust snow observations that are capable of being used unequivocally. The End. The paper leaves it to the reader to decide where they think we are currently placed in this four act play.

[1] University of Waterloo, Waterloo, ON, Canada

Global Snow from Space: Development of a Satellite-based, Terrestrial Snow Mission Planning Tool

BARTON FORMAN[1], YONGHWAN KWON[1], SUJAY KUMAR[2], YEOSANG YOON[2],
JACQUELINE LE MOIGNE[2], MATTHEW HOLLAND[2], AND SREEJA NAG[2,3]

ABSTRACT

A global, satellite-based, terrestrial snow mission planning tool is proposed to help inform experimental mission design with relevance to snow depth and snow water equivalent (SWE). The idea leverages the capabilities of NASA's Land Information System (LIS) and the Tradespace Analysis Tool for Constellations (TAT-C) to harness the information content of Earth science mission data across a suite of hypothetical sensor designs, orbital configurations, data assimilation algorithms, and optimization and uncertainty techniques, including cost estimates and risk assessments of each hypothetical permutation.

One objective of the proposed observing system simulation experiment (OSSE) is to assess the complementary – or perhaps contradictory – information content derived from the simultaneous collection of passive microwave (radiometer), active microwave (radar), and LIDAR observations from space-based platforms. The integrated system will enable a true end-to-end OSSE that can help quantify the value of observations based on their utility towards both scientific research and applications as well as to better guide future mission design. Science and mission planning questions addressed as part of this concept include:

1. What observational records are needed (in space and time) to maximize terrestrial snow experimental utility?

2. How might observation be coordinated (in space and time) to maximize this utility?

3. What is the additional utility associated with an additional observation?

4. How can future mission costs be minimized while ensuring Science requirements are fulfilled?

[1] University of Maryland, College Park, MD, USA
[2] NASA Goddard Space Flight Center, Greenbelt, MD, USA
[3] Bay Area Environmental Research Institute, Petaluma, CA, USA

Noisy Data or Noisy Landscape? Putting New Calibrated, Enhanced-Resolution Brightness Temperatures to the Test

JOAN M. RAMAGE[1], MITCHELL T. JOHNSON[1], MARY J. BRODZIK[2], TARA J. TROY[1], MOLLY A. HARDMAN[2], AND DAVID G. LONG[3]

ABSTRACT

With funding from the NASA MEaSUREs program, the National Snow and Ice Data Center (NSIDC) produced the Calibrated, Enhanced-Resolution Brightness Temperatures (CETB) Earth System Data Record (ESDR) for the complete record of legacy and ongoing SSM/I, SSMIS, AMSRE (and soon SMAP) sensors. CETB data were created using the radiometer version of the scatterometer image reconstruction (rSIR) technique. These enhanced-resolution data are 64 times higher spatial resolution (3.125 km pixels at 36/37 GHz frequencies) and 8 times high spatial resolution (6.125 km pixels at 18/19 GHZ frequencies) than the historical 25 km data products. They appear to provide significant improvement in the ability to distinguish finer spatial patterns. We started to work with them for assessing snow melt timing and snow water equivalent in diverse and heterogeneous landscapes. Colleagues asked us "How do you know your data are that good? Aren't they just noisier?" So, we set out to test whether the higher spatial resolution successfully captures accurate differences in real, heterogeneous landscapes. Sites are selected at or near locations with ground-based observations to aid in interpretation and understanding of detected variations. We compare sites with similar characteristics, controlling for variation in topographic relief, standing water, and land cover. Sites are compared using average, minimum and maximum brightness temperature, diurnal variability, and intra-pixel variability for pixels at differing resolutions compared to elevation and other factors. We found that CETB data, though having a slightly higher noise level than conventional products, also have finer effective resolution that better matches high resolution comparison sets.

---

[1] Lehigh University, Bethlehem, PA, USA

[2] National Snow and Ice Data Center, Boulder, CO, USA

[3] Brigham Young University, Provo, UT, USA

75th EASTERN SNOW CONFERENCE
College Park, Maryland, USA 2018

Assessment of the Stability of a Satellite-Derived Snow Extent Using Station Snow Depth Observations

DAVID A. ROBINSON[1] AND THOMAS L. MOTE[2]

ABSTRACT

Station and satellite-based measures of snow cover extent produce complementary, but occasionally conflicting, climate data records (CDRs). A record of snow cover extent from visible satellite data is available on a weekly basis since the late 1960s, while a gridded station product for North America is available on a daily basis for the past century. We have documented several changes in the visible satellite record. The effects of these discontinuities in the satellite CDR — such as different sensors, overpass times, or mapping methodologies — have not been fully assessed. Here the gridded North American snow depth record is compared to the visible satellite record for different epochs within the satellite CDR. The average snow depths in the station product are identified for the 50th percentile probability of snow identification in the satellite product. The dataset is broken into three different time periods, 1965 to 1980 (Period 1), 1981 to 1998 (Period 2), and 1999 to 2009 (Period 3), corresponding roughly to periods of technological and processing changes in the NOAA snow charts used to generate the CDR. A logistic regression model is employed to determine the probability of snow cover detection in any satellite cell based on two variables: average snow depth reported from weather stations, and fraction of reporting stations reporting at least 1 cm of snow. The modeled results are used to adjust the earlier two periods based on the most recent period, and the adjustment indicates that most recent period shows 5.23% greater snow extent than Period 1, and 2.75% more than Period 2.

[1] Rutgers University, Piscataway, NJ, USA
[2] University of Georgia, Athens, GA, USA

Session #2: In-Situ Snow Measurements & Field Experiments

Snow Water Equivalent from Operational GNSS In-situ Stations as Service for Hydrological Applications ESA IAP Demo in Eastern Canada

FLORIAN APPEL[1], FRANZISKA KOCH[2], AND PATRICK HENKEL[3]

ABSTRACT

Especially in remote areas, e.g. in Canada or Scandinavia, the density of SWE measurements is up to now still very limited and all operations (stations and field trips) are usually costly, labour-intense and/or risky.

Based on a novel GNSS snow measurement approach, the team designed, developed and demonstrated an operational service for the island of Newfoundland/Canada within the European Space Agency ESA co-founded demonstration project SnowSense (2015-2018). The service is based on a system including autonomous GNSS based snow monitoring stations, Earth Observations (EO) of the snow cover, and an integrated hydrological model component (PROMET). This combined system approach enables the provision of spatial SWE information, run-off assessment and relevant information for hydropower plant management, particularly for so far non- or sparsely equipped catchments in remote areas. The key element of the system is the novel GNSS based in-situ sensor, using static low-cost antennas and receivers, on the ground and on above the snow. These sensors are able to retrieve the snow parameters SWE and LWC (liquid water content in snow), using carrier phases and carrier-to-noise power density ratio measurements, and send the information via satellite communication several times a day. In combination with EO and the GNSS in-situ stations, the spatially results of the model are controlled and updated to provide all hydrological parameters, like the water stored in the snow cover and the river run-off.

The authors will present the latest results of the station design, the performance, the snow measurements and the hydrological products, demonstrated within the demo operation 2017/2018 for Newfoundland. In-situ stations are also in operation in Quebec/Canada, Germany and Switzerland. A commercial service roll-out is the overall target for the next winter.

STATION DESIGN & OPERATION

Focus of the development and design was to provide a solution for the retrieval of SWE measurements at locations so far not equipped with stations or not regularly visited by field trips.

Based on the proven technology of using GNSS/GPS signals for snow parameter, a transportable, but also modular to be integrated in existing stations, solution has been designed, produced and

[1] VISTA Remote Sensing in Geosciences GmbH, Munich, Germany
[2] Department of Geography, Ludwig-Maximilians-Universität, Munich, Germany
[3] ANavS GmbH, Munich, Germany

tested. Using the own power and communication sources, the in-situ stations can provide snow parameter on a daily basis, without geographical restrictions.

Stations, primary developed for application in Newfoundland, had been additionally tested in Austria, Germany and Switzerland. First devices had been sold to national European science institutes and authorities. Station design and method of measurement are patented.

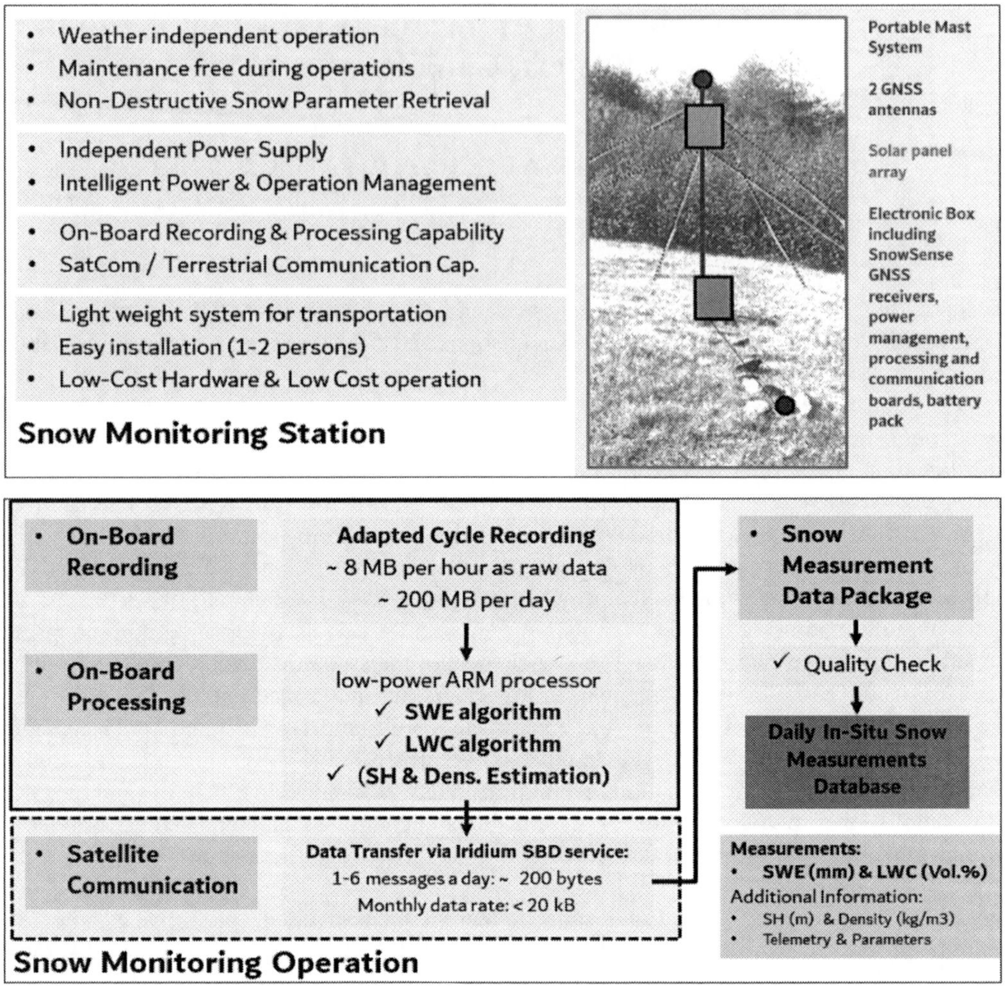

Figure 1. Requirements and features of the *in-situ* station at a glance. Design and operation is dedicated to flexible installation and independent operation in remote areas.

STATION PERFORMANCE IN QUEBEC/CANADA

Beside the installations in Newfoundland (8 stations under operation in Winter 2017/18), a demo station was installed at the Forêt Montmorency site in Quebec *, in cooperation with the University Sherbrooke and University Laval. At this location numerous meteorological instruments are in operation. Regular manual measurements complete the information.

The SnowSense station was installed in October 2017 and delivered daily SWE and LWC measurements without significant discontinuities. Power and communication system resisted the cold and wind environment without damage or failure. For a complete analysis, the recorded raw datasets had been additionally re-processed.

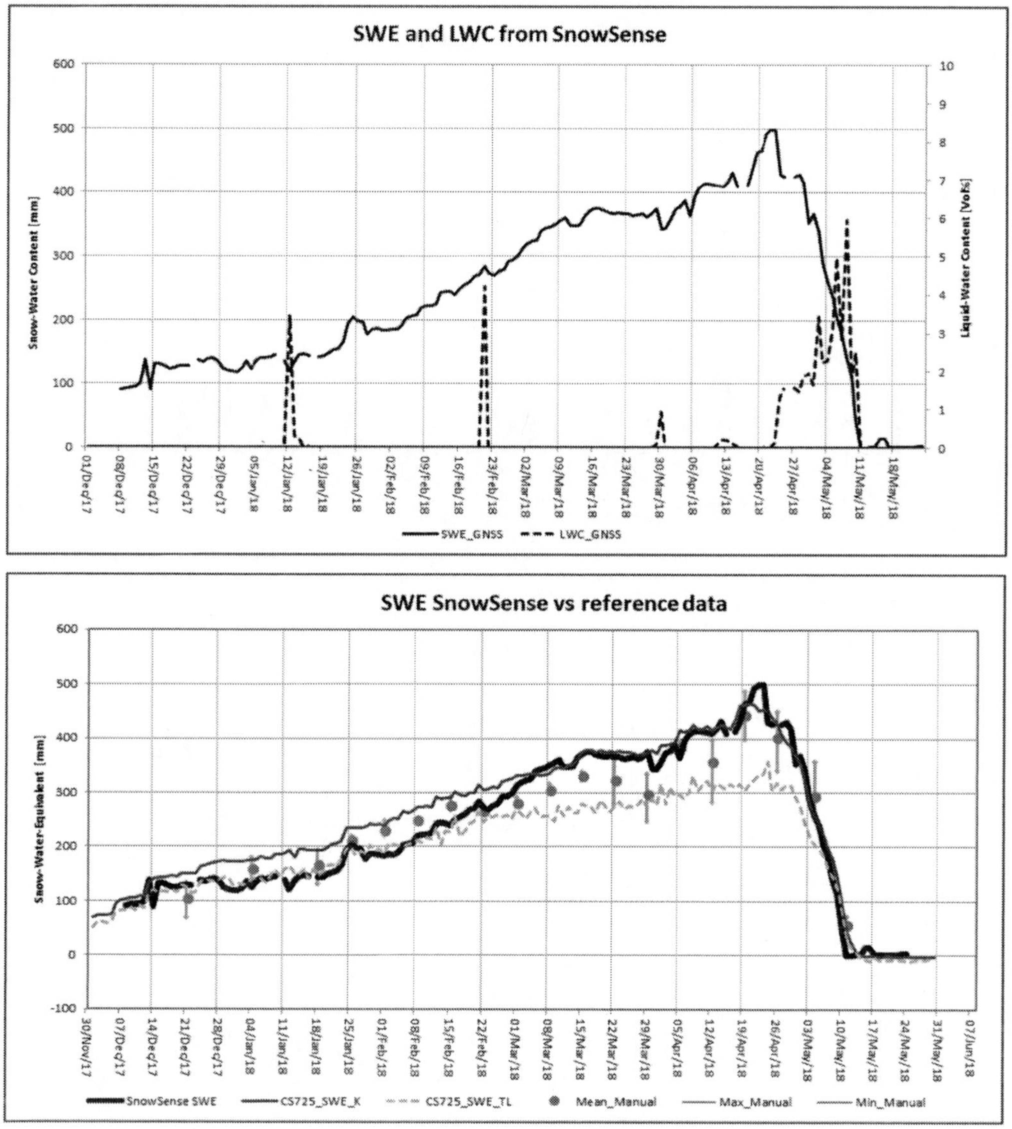

Figure 2. Upper panel: Daily SWE and LWC values at the Momo test site 2017/2018. Lower panel: Daily SWE and reference measurements (Campbell CS725 and manual probes).

The received SWE values are in good concordance with the provided comparing / reference measurements by a Campbell CS725 sensor and manual measurements. Differences / divergences between the measurement algorithms are under investigation. All measurements are from the same site, but different footprint. All preliminary reference data by courtesy of Univ. Laval and Univ. Sherbrook – detailed publications are planned.

In addition to the SWE, the SnowSense Sensor can provide measurements of the Liquid Water Content of the snow pack. With the LWC (measured several times a day) a good correspondence with rainfall and warm temperature events could be observed during the winter.

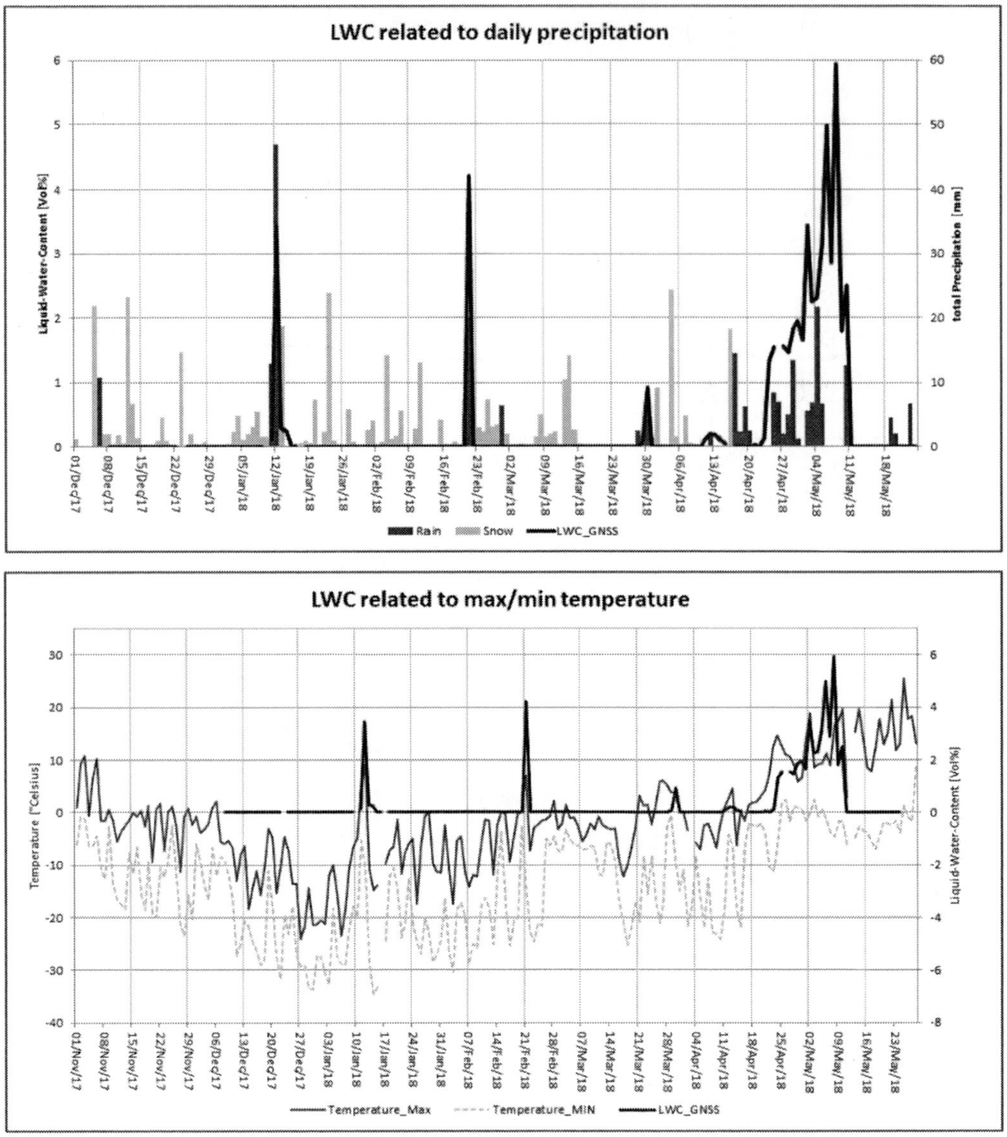

Figure 3. Daily LWC values at the Momo test site 2017/2018 related to rainfall events (upper panel) and warm temperature occurrence (lower panel). (Temperature and precipitation measurements received from the close by provincial and federal station site **.)

SERVICE PERFROMANCE IN NEWFOUNDLAND/CANADA

In addition to local measurements, SnowSense demonstrated it capabilities in providing spatial snow information and calculation of the run-off for the entire island of Newfoundland (~100.00 km^2). Based on the static information on the terrain (DEM, land use, soil) and the hydraulic situation (rivers, lakes, reservoirs) the daily product (had been released. Beyond maps of the SWE, the run-off Q was derived for several river locations in the main catchments. Based on the concept and design of SnowSense Service, the calculations had been forced by HRDPS*** data only (no meteorological station measurements, which are hardly available), assimilated with the information from online in-situ SWE information. Beyond the SWE and LWC information on a daily basis, EO observations, using Sentinel-1 satellites, had been used to control and adapt the model calculations. Regarding the run-off formation, basic information on the river network, the lakes and of single reservoirs / managed lakes, had been included. Due to the large area and the demo purpose of the service for the winter season, the parametrization of the catchment characteristics was not performed to the final level of details.

Results of the runoff calculation had been validated against available measurements from the Water Resources Management Divisions near-real-time stations ****, which are provided without quality checks. For the catchments of Humber River where a strong interest in flood forecast is predominating, the runoff results at one location are presented.

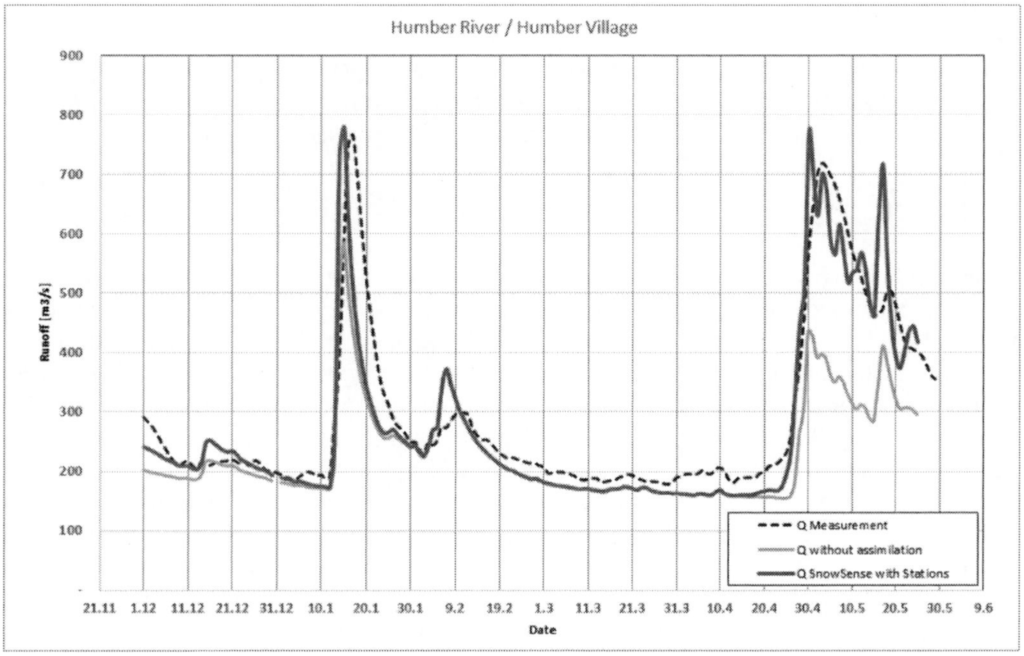

Figure 4. Daily runoff results from SnowSense/PROMET without (grey), with assimilation (blue) of stations and EO observations against the values measured at Humber Village (catchment: ~7600 km^2).

REFERENCES

Henkel P, Koch F, Appel F, Bach H, Prasch M, Schmid L, Schweizer J, Mauser W. 2018. Snow water equivalent of dry snow derived from GNSS carrier phases. *IEEE Transactions on Geoscience and Remote Sensing* **56**(6), 3561-3572. doi: 10.1109/TGRS.2018.2802494

Koch F, Henkel P, Appel F, Schmid L, Bach H, Prasch M, Schweizer J, Mauser W. In prep. Retrieval of water equivalent of dry and wet snow by combining GPS time delay and signal attenuation.

Koch F. 2017. *Snow cover properties and soil moisture derived from GPS signals*. Dissertation at the Faculty of Geosciences, Ludwig-Maximilians-Universität München. 143 pp. Available online: https://edoc.ub.uni-muenchen.de/21938/

Koch F, Prasch M, Schmid l, Schweizer J, Mauser J. 2014. Measuring snow liquid water content with low-cost GPS receivers. *Sensors* **14**: 20975-20999. doi: 10.3390/s141120975

* http://www.cef-cfr.ca/index.php?n=Membres.SylvainJutrasNEIGE

** http://climat.meteo.gc.ca/climate_data/hourly_data_e.html?StationID=32413

*** https://weather.gc.ca/grib/grib2_HRDPS_HR_e.html

**** http://www.env.gov.nl.ca/wrmd/ADRS/v6/Graphs_List.asp

http://www.vista-geo.de/en/snowsense

https://www.facebook.com/SnowSenseDemo/

2018 Hubbard Brook Field Experiment: Snow Observations in a North-Eastern U.S. Forested Region

CARRIE VUYOVICH[1], ALEXANDRE LANGLOIS[2], ALEXANDRE ROY[2], THEODORE LETCHER[1], JENNIFER JACOBS[3], JULIE PARNO[1], RONNY SCHROEDER[3], SIMON KRAATZ[3], ZOE COURVILLE[1], AND EUNSANG CHO[3]

ABSTRACT

Space-borne passive microwave observations of snow are impacted by forest cover. Various efforts have focused on improving snow characterization in forested regions to better understand the signal impacts. In February 2018, a two-day field experiment was conducted to collect ground-based radiometer and snow observations under various canopy types in the Northeast U.S., an under-studied region for this application. Nine sites were selected in the Hubbard Brook Experimental Forest (HBEF) with various canopy types and topographic properties. At each site, a ground-based radiometer was used to measure snow and vegetation brightness temperatures at 10, 19, 37, and 89 GHz. Snow pits within the footprint of the radiometers were measured to characterize the snowpack profile. Measurements included depth, density, temperature, liquid water content and grain size or specific surface area (by visual identification and short-wave infrared reflection method). Transects of approximately 30 m were conducted to measure representative snow depth and SWE at each of the sites. Photo of the canopy with a fish-eye lens (leaf area index: LAI) and albedo data were also collected at each location. Additionally, at selected locations, snow casts were collected and analyzed using a micro-CT scanner to determine microstructural characteristics. Preliminary results from this field campaign are presented. These data are expected to improve our understanding of appropriate parameters for use in microwave snow radiative transfer and microstructure models for this region.

---

[1] U.S. Army Engineer Research and Development Center, Hanover, NH, USA
[2] Université de Sherbrooke, Sherbrooke, QC, Canada
[3] University of New Hampshire, Durham, NH, USA

Snowy Opportunities at the NEIGE Site, Montmorency Forest, Québec, Canada

AMANDINE PIERRE[1] AND SYLVAIN JUTRAS[1]

ABSTRACT

Although it is recognized that under-catch is a significant source of bias in the measurement of solid precipitation and snow water equivalent (SWE), this factor is still not standardized within monitoring networks in Canada. This situation is a source of considerable uncertainties, among others, when snow is an essential input for the simulation of river flows.

Few sites around the world can enable the inter-comparison of precipitation gauges and SWE protocols, but some of them are found in Canada. The NEIGE site, located in Montmorency Forest, Québec, Canada, is currently the most important multi-institutional experimental site for solid precipitation studies in Québec and among the most equipped in Canada.

Numerous partners are involved in the development, since 2014, of this experimental site dedicated to snow research. Located in a very snowy environment (mean annual snowfall of 619 mm), this easily accessible site enables the measurement of unshielded and shielded (Double Fence Intercomparison Reference [DFIR], Bush, Nipher, Alter, Double-Alter, Tretiakov) gages (mass and volume manual recipients, Pluvio² OTT, Geonor T200B). The continuous availability of competent bi-daily meteorological observers is also an irreplaceable benefit of this site. Measurement of SWE (4 manual protocols, gamma sensors [CS725/GMON], microwave radars, high resolution GPS), phase (Parsivels OTT), snow depth (manual scale and SR50) and many more meteorological parameters, are enabling the development and validation of transfer functions and snow models. Research objectives, recent results and upcoming opportunities rising from this unique site will be presented.

[1] Université Laval, Québec, QC, Canada

History of Ice on the Chesapeake Bay

JAMES L. FOSTER[1]

EXTENDED ABSTRACT

Keywords: Chesapeake Bay, ice cover, severe winter

INTRODUCTION

Ice forms in most winters, at least for a few days, in the colder, fresher waters of the upper Chesapeake Bay. As a rule of thumb, ice in the upper-Bay is a concern to navigation and commercial fishing about every 6 or 7 years. Chesapeake Bay ice is of course more noteworthy and more widely reported following a period of consecutive years during which time the Bay's water had generally been ice free – between the winter of 1917-18 and 1935-36 and 1982-83 and 2015, for example.

Because the average air temperature in the Bay area for January and February is approximately 0 °C, prolonged cold is required to ice over the mid and lower-Bays. It should be noted that even in relatively warm winters, ice will form if a blast of polar air cools the surface water to the freezing level, but in these instances the ice cover is more of a veneer and not a serious impediment to watermen or to navigation. On the other hand, in winters when air temperatures are below normal but winds are persistent, ice formation may be delayed. But if air temperatures during autumn are cool, and the Bay's waters are thus below normal entering winter, the average air temperature in January is a bellwether in determining how severe icing will become.

STUDY AREA

The Chesapeake's tidal shoreline (at high tide) stretches approximately 11,788 km; adding in the mileage of its tidal tributaries this total increases to 18,804 km. It takes in water from an area totaling nearly 165,729 km^2; from most all of Maryland, from about half of Virginia, Delaware and Pennsylvania, from about one fifth of West Virginia, New York, and even from a few square miles of North Carolina. The majority of the Bay's water volume, between 50-55 %, flows south from the Susquehanna River; another 20- 25% is contributed by the southeastwardly drainage of the Potomac River (Fisher, 1980).

RESULTS AND DISCUSSION

Figure 1 shows a timeline of biggest ice years in the Bay from 1600 to 2020. The years that ice extended into the lower Bay, for four weeks or more, are plotted in blue; whereas the years the ice extended into the mid Bay, for a period several weeks, are plotted in orange. In each century, at least two winters were sufficiently cold that nearly the entire Bay was locked in ice. Based on both anecdotal evidence (letter writing and newspaper articles) (Ludlum, 1966), and in recent decades on

[1] NASA Goddard Space Flight Center (Emeritus), Greenbelt, MD, USA

satellite observations, during these epic ice years the ice cover extended well into the lower Bay for a period of four weeks or more. Over the past 400 years, there appears to be no obvious trend toward either a greater separation or to a smaller separation in the number of years between epic ice events.

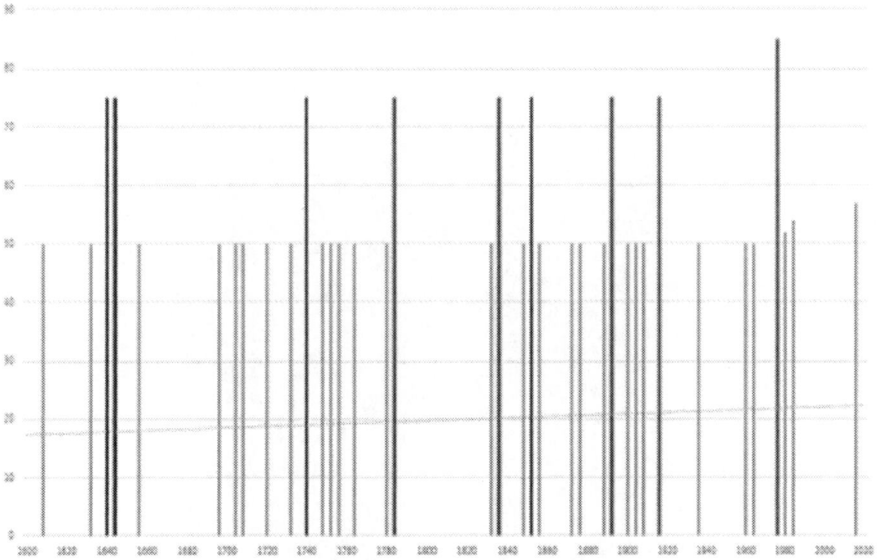

Figure 1. Big ice years on the Chesapeake Bay, 1600-2018. Ice extent in percent (y-axis).

During years when ice extent was much above normal, even during the ten extraordinary years when the ice was very extensive, air temperatures in early spring following melt, on either the Western Shore or the Eastern Shore did not seem to be affected by the icy waters. At Baltimore, for example, there was virtually no correspondence between maximum winter ice extent and air temperature the following March. In fact, for the eight years since 1900 having the most severe ice, only in March of 1904 was the air temperature more than 1 degree C below normal. In essence it seems that after an exceptionally cold winter the atmosphere is spent; the jet stream migrates to the north shutting off cold air to the Chesapeake Bay region.

Therefore, it should not be surprising that there is little persistence between winter and summer temperatures; a cold winter does not portent a pleasant summer. In 1918, nearly the entire Bay was frozen in January, but the temperature soared to record levels that summer, reaching 41 °C in Baltimore in August. Moreover, though the winter of 1936 was bone-chilling, record-breaking heat was in place in July and August (Dawson, 2008).

Nor is there an obvious association between the severity of the cold or the extent of the ice in a given winter and conditions either the winter before or the winter following. In 1635, for instance, it was mentioned in letters that "this last winter was the coldest that has been knowne in many yeeres: but the yeere before, there was scarce any signe of Winter, onely that the leaves fell from the trees, in all other things it appeared to be Summer."

January is generally the month having the greatest ice extent; in the biggest ice years and also during years when the ice cover is not severe. But because of the thermal inertia of water, the ice cover can be extensive in late winter (February and March) even when air temperatures in February are above normal. This happened in February of 1977, for instance. In several of the big ice years, but probably not in any of the epic ice winters, the maximum ice extent occurred in February, for example in February 1895, 1899, 1978, 1979 and in 2015.

Figure 2 shows ice extent on the Chesapeake Bay (middle and lower Bay) as observed from Landsat on February 2, 1977.

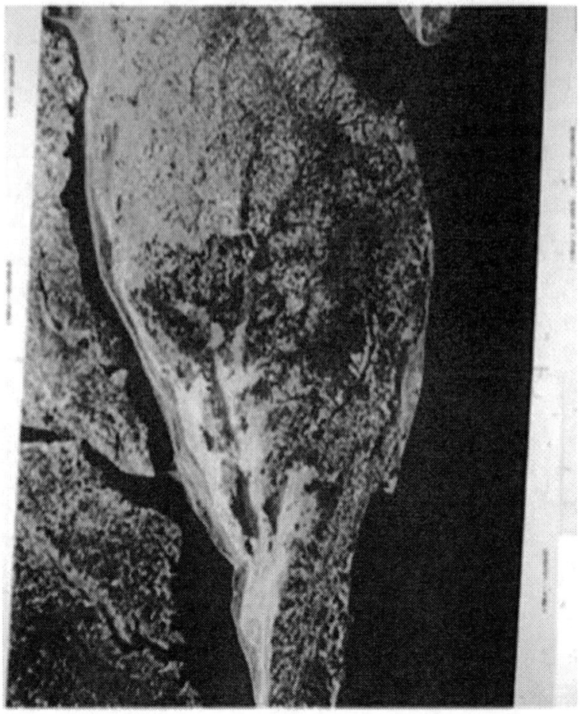

Figure 2. Ice cover (blue) in the middle and lower Chesapeake Bay in Early February 1977.

SUMMARY

Even though air temperatures have noticeably warmed in recent decades as a result of climate change, in most every winter there is a period of a few days to a few weeks when the average air temperature remains below freezing (the freezing degrees days are positive). When this occurs ice forms, if only in the upper Bay and if only for a week or less. From time to time, however, cold Arctic air pours across the upper Mid-West and into the eastern U.S., nearly uninterrupted for weeks at a time. These are the winters that ice cover in the Chesapeake Bay becomes remarkable. If autumn temperatures happen to be below normal, so that the Bay water is pre-chilled, icing will be truly extraordinary if the winter is exceptionally cold.

REFERENCES

Dawson J. 2008. History of Maryland Weather including Chesapeake Bay. www.unicordbookshop.com/pdf/md-weather-summary.pdf.

Fisher Jr. AC. 1980. My Chesapeake – Queen of Bays. National Geographic Magazine, **158**(4).

Ludlum DL. 1966. Early American Winters, 1604-1820. American Meteorological Society, Boston, MA: 162-163.

Session #3: Snow Modeling & Snow Processes

A New Estimate of North American Montane Snow Water Equivalent: Validation Challenges and Large-Scale Implications

MELISSA L. WRZESIEN[1,2], MICHAEL T. DURAND[1,2], TAMLIN M. PAVELSKY[3], SARAH KAPNICK[4], YU ZHANG[1], JUNYI GUO[1], AND C. K. SHUM[1,5]

ABSTRACT

Despite the importance of seasonal snow to the hydrologic cycle, global mountain snow storage estimates are highly uncertain. Observation networks in mountainous regions are sparse, and satellite retrievals can perform poorly in mountains. Models are one of the few options for estimating snow over large geographical areas, such as entire mountain ranges or entire continents. Here we use a regional climate model to produce a new estimate of mountain seasonal snow accumulation for North America. From this work, we suggest there is 1006 km^3 of snow water storage (SWS) across the continent's mountains, which nearly three times more SWS than previous estimates. Over the entire continent, we estimate a peak SWS of 1684 km^3, 55% greater than previous estimates. However, a larger challenge is evaluating our new climatological estimate. In situ measurements may not be representative of the surrounding area, which is problematic when comparing to a 9 km model grid cell. Evaluating against other models is difficult since every model estimate has its own uncertainties; additionally, previous work suggests that many global data products underestimate mountain snow. Nevertheless, we evaluate our North American SWS dataset against in situ observations from snow pillows (bias of -89 mm), remotely sensed snow cover fraction, model-estimated snow water equivalent (bias of +12 mm, compared to SNODAS), and terrestrial water storage anomalies from GRACE. Though a formal model validation is impossible, from the comparisons presented here, we are able to determine whether our estimate is reasonable for SWS. Perhaps more importantly, from these evaluations, we are able to consider mountain seasonal snow accumulation in the broader perspective of the entire continental water budget.

[1] School of Earth Sciences, The Ohio State University, Columbus, OH, USA
[2] Byrd Polar and Climate Research Centre, The Ohio State University, Columbus, OH, USA
[3] Department of Geological Sciences, University of North Carolina, Chapel Hill, NC, USA
[4] Geophysical Fluid Dynamics Laboratory, National Oceanic and Atmospheric Administration, Princeton, NJ, USA
[5] Institute of Geodesy & Geophysics, Chinese Academy of Sciences, Wuhan, China

Impacts of Vegetation and Snow on Permafrost Variability

JING TAO[1], ROLF H. REICHLE[2], RANDAL D. KOSTER[2], BARTON A. FORMAN[3], AND YUAN XUE[4]

ABSTRACT

Vegetation plays a critical role in modulating snow accumulation processes. Snowpack, acting as an insolation layer (i.e., "thermal blanket"), impedes heat exchange between ground and atmosphere and thus affecting subsurface thermodynamics. In this study, we use the NASA Catchment Land Surface Model (CLSM) driven by the Modern-Era Retrospective Analysis for Research and Applications-2 (MERRA-2) land forcing fields to simulate active layer thickness (ALT) over permafrost regions in the Northern Hemisphere. We first demonstrate that most of the ALT variability can be jointly explained by accumulated air temperature and maximum snow water equivalent (SWE) in the CLSM-identified permafrost regions. Then, we discuss the impacts of vegetation and snow on ALT at several locations in high-latitude permafrost regions. At one particular site in Alaska, we show that replacing vegetation cover in the CLSM with the local vegetation type leads to improvements in the simulation results of snow depth, soil temperature profile, and ALT. Sensitivity analysis reveals that a thicker snowpack in winter season is able to facilitate a deeper ALT later in the warm season. That is, a larger snow depth could better slow down the heat release from soil to the atmosphere during the cold season, causing a warmer subsurface soil temperature and then a deeper thaw depth in summer. At last, we explore realistic methods to improve model simulation results.

[1] Earth System Science Interdisciplinary Center, University of Maryland, College Park, MD, USA
[2] NASA Goddard Space Flight Center, Greenbelt, MD, USA
[3] Department of Civil and Environmental Engineering, University of Maryland, College Park, MD, USA
[4] George Mason University, Fairfax, Virginia, USA

Impact of Heat Convection Induced by Topography-Driven Air Ventilation on Snow Surface Temperature

NICOLAS R. LEROUX[1], JOHN W. POMEROY[1], AND WARREN D. HELGASON[1,2]

INTRODUCTION

Exchange of energy between the atmosphere and the snowpack drives most snowmelt (Male and Granger, 1981; Gray and Landine, 1988). The snow-atmosphere boundary is a permeable wall that permits penetration of wind flow. Air ventilation in snow and firn has been observed in both natural and artificial snowpacks (Albert and Hardy, 1995; Clifton et al., 2008; Drake et al., 2017). Airflow through pore spaces impacts the thermal regime of ice sheets and snowpacks by transporting heat and water vapour (Clarke et al., 1987; Powers et al., 1985), as well as the transport of chemical species (Waddington and Cunningham, 1996). Albert and Hardy (1995) conducted a field experiment to induce wind flow at the surface of a snowpack whilst measuring temperature distributions within the snowpack. They observed highly non-uniform internal snow temperature distributions.

Some models have represented snow-air exchange by 2D advection-conduction heat transfer algorithms (Albert and Hardy, 1995) or other schemes (Colbeck, 1989, 1997; Clarke and Waddington, 1991, Albert, 1993). However, in most numerical models of snow-atmosphere exchange, assumptions are made to simplify the energy balance; for instance, turbulent fluxes are represented by 1st order theories and the snow surface is considered an impermeable and flat boundary. Some studies have used the radiative snow surface temperature as an index to assess snow model performance (Lapo et al., 2015; Conway et al., 2018). This index has exposed the under-estimation of turbulent fluxes by the current theories under stable atmospheric conditions that are often encountered above snow (Cullen et al., 2007). In particular, Helgason and Pomeroy (2012) observed that the measured sensible flux could not offset heat loss through longwave radiation during clear sky conditions, which resulted in unrealistic drops in simulated snow surface temperature using the SNTHERM model. Helgason and Pomeroy (2012) suspected that forced convection into the snowpack under the presence of sastrugi might have influenced measured snow surface temperature. To resolve this issue in models, a numerical strategy of adding a 'windless coefficient' to increase estimated sensible heat flux under stable conditions is often employed (Brun et al., 1989; Jordan et al., 1999; Brown et al., 2006). This method, however, lacks physical realism.

This study investigates the impact of thermal convection induced by topography-driven pressure fluctuations at the snow surface coupled with surface energy balance on simulated near snow surface temperature.

[1] Centre for Hydrology, University of Saskatchewan, Saskatoon, SK, Canada
[2] Department of Civil, Geological, and Environmental Engineering, University of Saskatchewan, SK, Canada

METHOD

Snow Energy Balance

Pomeroy et al. (2016) developed an energy balance model to solve for the radiative snow surface temperature, assuming that the snow surface is thermally disconnected from the underlying snow (i.e. there is no heat conduction between the snow surface and the underlying layer). Their model considered a no-mass "snow surface skin" of zero internal energy. In this study, the energy balance at the snow surface is applied to an exchange layer of a few centimetres. This exchange layer is also assumed to have no mass, but in contrast to Pomeroy et al.'s model, is linked to the snowpack via a heat conduction term.

A first-order closure was applied to estimate the vertical fluxes of momentum, sensible heat, and latent heat, i.e. that these fluxes are proportional to the vertical gradients of wind speed, temperature, and specific humidity, respectively. To account for atmospheric stability, the Monin-Obukohv similarity theory is used (Monin and Obukhov, 1954).

Thermal Convection in Porous Media

The topography-driven airflow theory is taken from previous studies of airflow through snow (Bartlett and Lehning, 2011). The velocity of the air phase was estimated using Darcy's law. The pressure at the snow surface was divided into a stationary, homogeneous ambient pressure and a time-varying, inhomogeneous pressure fluctuation of smaller magnitude. The pressure fluctuation at the snow surface depends on the length and height of snow dunes. For simplicity, a flat surface was approximated, following Bartlett and Lehning (2011).

Heat Transfer

Liquid water within the pore space was neglected. Thus, the snowpack is only composed of ice and air. Assuming incompressible airflow and thermal equilibrium between the gas and solid phases, heat transfer through snow was estimated using the convection-conduction heat equation.

Model Design

A 2D model was developed to simulate the energy balance near the snow surface and heat transfer through advection and conduction within the snowpack. An explicit upwind finite volume method was applied to solve for the 2D heat convection-conduction equation. The snowpack temperature profile was initially assumed to be linear, ranging from -10oC at the surface to 0oC at the bottom. A constant temperature was assumed at the bottom of the snowpack (=0oC) and periodic boundary conditions were specified at the lateral boundaries.

The meteorological inputs were assumed constant and are summarized in Table 1.

RESULTS

Impact of Dune Height

To investigate the impact of the height of snow dunes on simulated snow surface and internal temperature, the wavelength of the snow dune was kept constant at 2 m. Snow properties, domain boundaries, and atmospheric inputs for the reference case are summarized in Table 1. The height of the snow dune was increased from 5 cm to 15 cm by increments of 5 cm for two different air temperatures (-20 °C and -10 °C). All other variables were held constant and the simulations were run until steady state conditions developed. Figure 1a presents the lateral (horizontal) distribution of the simulated snow surface temperature for three different dune heights, at an air temperature of -20 °C. The snow surface temperature was not homogeneous. At each dune height, the location of maximum snow surface temperature corresponded to the location of minimum pressure at the snow surface. Inversely, the lowest snow surface temperature was estimated where the pressure at the surface was maximal (i.e. where the air moved downward within the snowpack). For increasing dune height, the maximum value of the snow surface temperature increased, while the minimum value remained similar at all dune heights. The difference between the maximum and minimum values of the estimated snow surface temperatures over the snow surface (ΔT_s), which symbolizes the heterogeneity in the lateral distribution of snow surface temperature, for all three dune heights and at the two different air temperatures, are shown in Fig.1. At all dune heights, ΔT_s decreased with increasing air temperature.

Table 1. Snow properties, model boundaries, and atmospheric inputs for the reference case.

Property	Value
Density	300 kg m^{-3}
Grain size	1 mm
Temperature at snow-soil interface	0 °C
Initial snow surface temperature	-10 °C
Dune height	5 cm
Dune wavelength	2 m
Roughness length	1 mm
Height of measurement	2 m
Air temperature	-20 °C
Wind speed	5 m s^{-1}
Relative humidity	80%
Incoming shortwave radiation	100 W m^{-2}
Incoming longwave radiation	200 W m^{-2}

Heterogeneity in snow surface temperature can be explained by looking at internal snow temperature distribution (Fig. 2). For taller dunes, airflow within the snow was faster, resulting in greater thermal convection flux. This caused the cold front on the left-hand side of the snowpack (where the air flowed downward from the surface) to be colder and reach deeper snowpack layers. Conversely, the velocity of the upward airflow (on the right-hand side of the snowpack) increased with dune height and more thermal energy was transported from the warm ground-snow interface to the snow surface, resulting in a warmer internal snow temperature just below the surface on the right-hand side of the snowpack. This corresponded with higher snow surface temperatures simulated on the right-hand side of the snowpack for increasing dune height.

A similar analysis was conducted for two different wind speeds (5 m s^{-1} and 10 m s^{-1}) whilst the other variables were kept constant (Table 1). The new results were similar to the previous observations. The minimum value of snow surface temperature varied very little with wind speed and dune height, and the highest value of snow surface temperature increased with dune height and agreed with the location of the smallest pressure at the snow surface. ΔT_s increased with wind speed and dune height (Fig. 1b).

Impact of Dune Wavelength

In this section, the impact of dune length on simulated snow temperature is studied. Simulations were first run for two different air temperatures (-10 °C and -20 °C) at a fixed dune height of 5 cm. All the other model inputs and initial conditions were kept constant (Table 1). The results are presented after steady state was reached.

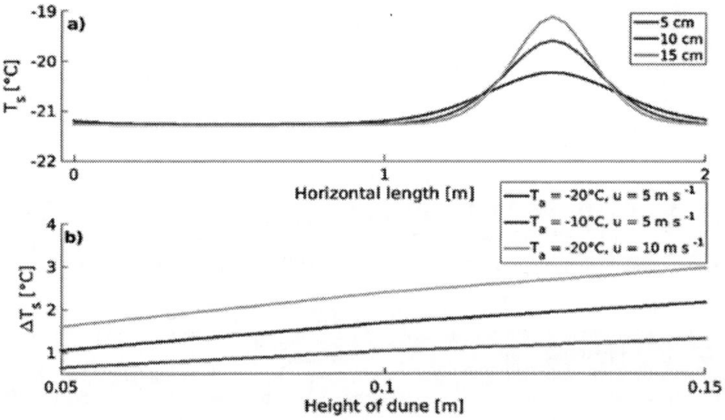

Figure 1. a) Horizontal distribution of simulated snow surface temperature for three dune heights (5 cm, 10 cm, and 15 cm) at an air temperature of -20 °C and a dune length of 2 m. b) Difference between the maximum and the minimum values of the distributed snow surface temperatures (dTs) for two different air temperatures (-20 °C and -10 °C) and increasing dune height(from 5 cm to 15 cm) with a constant dune length of 2 m.

Figure 3a shows the simulated lateral distribution of snow surface temperature for three different dune lengths (1 m, 5 m, and 10 m) and an air temperature of -20 °C. For longer dunes, the minimum value of snow surface temperature increased, while the maximum value decreased: snow surface temperature became more homogeneous. Figure 2b presents the variation of ΔT_s with air temperature and dune length. ΔT_s was greater for a colder air temperature. The distributed surface temperature flattened with increasing dune length; this was caused by lower air convection from a more uniform pressure gradient at the snow surface.

Wind speed was increased from 5 m s^{-1} to 10 m s^{-1} for a fixed air temperature of -20 °C. The variations of ΔT_s for these two wind speeds and the three different dune lengths are presented in Fig. 3b. Unexpectedly, ΔT_s was non-monotonic with dune length: at the highest wind speed, the maximum value of snow surface temperature was higher for a 5 m wavelength than for a 1 m and 10 m wavelength (Fig. 3b). This was caused by higher heat convection in 5 m wavelength dune, resulting in more transport of energy from the bottom of the snowpack (warm) to the snow surface.

CONCLUSIONS

The impact of thermal advection initiated by topography-induced airflow on snow internal and near surface temperatures was investigated. The input heat flux was estimated by solving for the energy balance at the snow surface, using meteorological data to drive the model. Increasing dune height resulted in an increase in heat convection and therefore in more spatial variability in estimated snow surface temperature. Heat convection within the snowpack decreased with increasing dune wavelength, resulting in a more homogeneous snow surface temperature. The near snow surface temperature was, however, more varied for a colder air temperature and a higher wind speed above the snow surface.

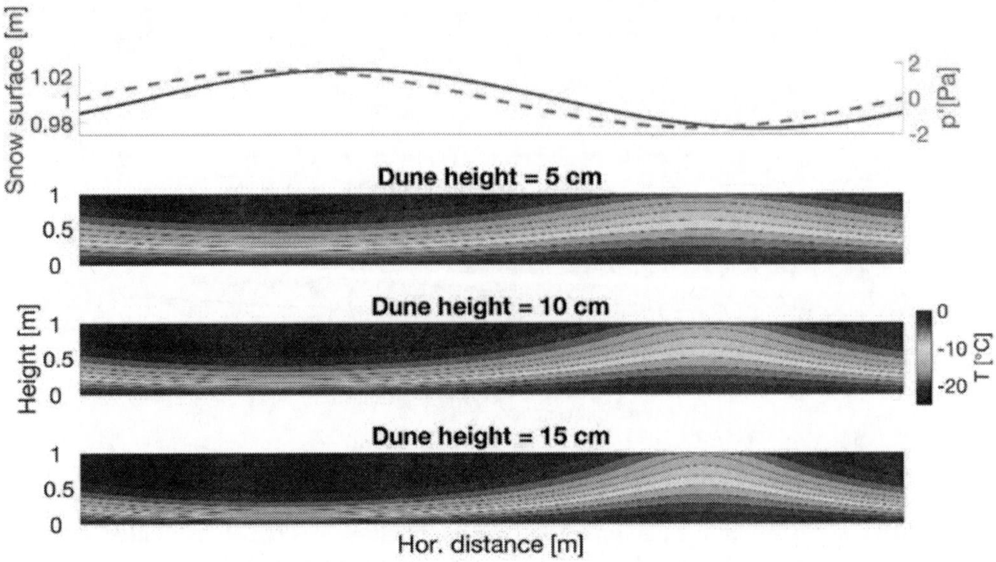

Figure 2. Simulated snow internal temperature with the values from the reference case (Table 1) for three different dune heights (5 cm, 10 cm, and 15 cm from top to bottom) and a constant dune length of 2 m. The top graph shows the topography of the snow surface and the pressure perturbation (p') generated by it.

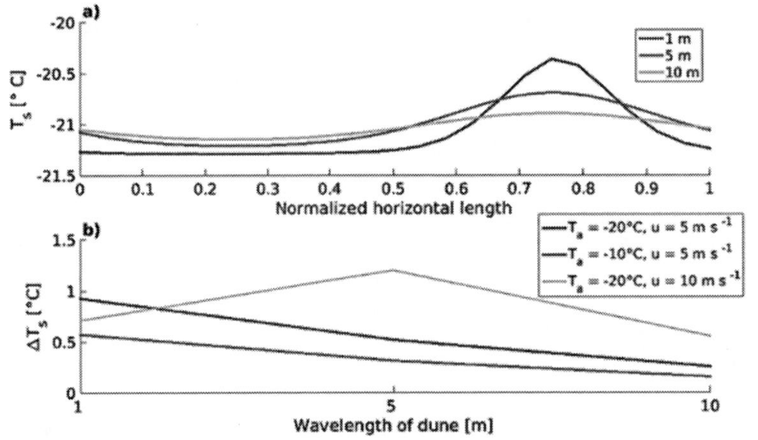

Figure 3. a) Horizontal distribution of snow surface temperature for three dune lengths (1 m, 5 m, and 10 m) and an air temperature of -20 °C, a wind speed of 5 m s^{-1}, and a dune height of 5 cm. b) Difference between the maximum and minimum values of the distributed snow surface temperature (ΔT_s) for two different air temperatures (-10 °C and -20 °C), two different wind speeds (5 m s^{-1} and 10 m s^{-1}) and an increasing dune wavelength (from 1 m to 10 m), with a fixed dune height of 5 cm.

ACKNOWLEDGEMENTS

The authors wish to acknowledge financial support from an NSERC Discovery grant, the NSERC Changing Cold Regions Network, the Canada Excellence Research Chair in Water Security, the Canada Research Chairs programme and the Global Water Futures initiative.

REFERENCES

Albert MR. 1993. Some numerical experiments on firn ventilation with heat transfer. *Annals of Glaciology*, **18**: 161-165. doi: 10.1017/S0260305500011435.

Albert MR, Hardy JP. 1995. *Ventilation experiments in a seasonal snow cover*. IAHS Publ. No. 228: 41–49.

Bartlett SJ, Lehning M. 2011. A theoretical assessment of heat transfer by ventilation in homogeneous snowpacks. *Water Resources Research*, **47**: 1–7. doi:11029/2010WR010008.

Brown R, Bartlett P, MacKay M, Verseghy D. 2006. Evaluation of snow cover in CLASS for SnowMIP. *Atmosphere-Ocean*, **44**: 223–238. doi:13137/ao.440302.

Brun E, Martin E, Simon V, Gendre C, Coleou, C. 1989, An energy and mass model of snow cover suitable for operational avalanche forecasting. *Journal of Glaciology*, **35**: 333–342.

Clarke GKC, Waddington ED. 1991. A three-dimensional theory of wind pumping. *Journal of Glaciology*, **37**: 89–96. doi:11017/S002214300004283.

Clarke GKC, Fisher DA, Waddington ED. 1987. *Wind pumping: A potentially significant heat source in ice sheets*. IAHS Publ. No. 170: 169–18.

Clifton A, Manes C, Rüedi JD, Guala M, Lehning M. 2008. On shear-driven ventilation of snow. *Boundary-Layer Meteorology*, **126**: 249–261. doi:10.1007/s10546-007-9235-0.

Colbeck SC. 1989. Air movement in snow due to windpumping. *Journal of Glaciology*, **35**: 209–213.

Colbeck SC. 1997. Model of wind pumping for layered snow. *Journal of Glaciology*, 43: 60–65.

Conway JP, Pomeroy JW, Helgason WD. 2018. Challenges in modelling turbulent heat fluxes to snowpacks in forest clearing. *Journal of Hydrometeorology*, in review.

Cullen NJ, Steffen K, Blanken PD. 2007. Nonstationarity of turbulent heat fluxes at Summit, Greenland. *Boundary-Layer Meteorology*, **122**(2): 439-455. doi:1007/s10546-006-9112-2.

Drake SA, Selker JS, Higgins CW. 2017. Wind enhances differential air advection in surface snow at sub-meter scales. *The Cryosphere*, **11**: 2075-2087. doi:15194/tc-11-2075-2017.

Gray DM, Landine PG. 1988. An energy-budget snowmelt model for the Canadian Prairies. *Canadian Journal of Earth Sciences*, **25**: 1292-1303.

Helgason W, Pomeroy JW. 2012. Problems closing the energy balance over a homogeneous snow cover during midwinter. *Journal of Hydrometeorology*, **13**: 557–572. doi:11175/JHM-D-11-0135.1.

Jordan RE, Andreas EL, Makshtas AP. 1999. Heat budget of snow-covered sea ice at North Pole 4. *Journal of Geophysical Research: Oceans*, 104: 7785–7806. doi:11029/1999JC900011.

Lapo KE, Hinkelman L, Raleigh M,.Lundquist JD. 2015. Impact of errors in the downwelling irradiances on simulations of snow water equivalent, snow surface temperature, and the snow energy balance. *Water Resources Research*, **51**: 1649–1670. doi:11002/2015WR01709.

Male DH, Granger RJ. 1981. Snow surface energy exchange. *Water Resources Research*, **17**: 609–627. doi:11029/WR017i003p00609.

Male DH, Gray DM. 1975. Problems in developing a physically based snowmelt model. *Canadian Journal of Civil Engineering*, **2**(4): 474–488.

Monin AS, Obukhov AM. 1954. Basic laws of turbulent mixing in the surface layer of the atmosphere. *Contributions of the Geophysical Institute, USSR Academy of Science*, **24**: 163–187.

Pomeroy JW, Essery RLH, Helgason WD. 2016. Aerodynamic and Radiative Controls on the Snow Surface Temperature. Journal of Hydrometeorology, **17**: 2175–2189. doi:11175/JHM-D-15-0221.

Pomeroy JW, Gray DM. 1995. *Snowcover Accumulation, Relocation and Management*. NHRI Science Report No. 7, Environment Canada and Minister of Supply and Services Canada: 1-135.
Waddington E, Cunningham J. 1996. The effect of snow ventilation on chemical concentrations. Wolff EW, Bales RG (eds.). *Chemical exchange between the atmosphere and polar snow*. NATO ASI series, **143**: 403–451.

Frequency and Timing of Snow Melt and Refreeze in the Northern U.S. from Satellite Brightness Temperature and Air Temperature

SAMUEL E. TUTTLE[1] AND JENNIFER M. JACOBS[2]

ABSTRACT

Knowledge of snow melt and refreeze events from satellite observations can be used to constrain snowpack metamorphism and stratigraphy, and runoff timing. Compared to an otherwise identical frozen snowpack, a snowpack that contains liquid water will emit more microwave radiation. Diurnal snow phase changes will thus lead to large changes in the brightness temperature (Tb) observed by a passive microwave radiometer, with higher brightness temperatures corresponding to wet snowpacks. We build on the diurnal amplitude variation (DAV) method of Ramage and Isacks (2002, Ann. Glaciol.) in order to identify individual snow melt and refreeze events. Here, we compare the difference between nighttime and daytime microwave Tb observations to coincident changes in air temperature (Ta). This allows the effect of diurnal snow phase changes on brightness temperature change to be isolated, by removing the effect of physical temperature change on brightness temperature change. Individual melt and freeze events are detected as large excursions from the modal linear regression line fit to the relationship between Tb change (ΔTb) and Ta change (ΔTa), using clustering techniques. This ΔTb-ΔTa method, previously validated at Senator Beck Study Basin, Colorado, is used examine the distribution of melt and refreeze events over the northern contiguous United States during the operational lifetime of the AMSR-E satellite instrument. We also examine some of the limitations of this method, including the difficulty of detecting snow melt and refreeze beneath vegetation canopy.

[1] Mount Holyoke College, South Hadley, MA, USA
[2] University of New Hampshire, Durham, NH, USA

Support Vector Machine Predictions of Passive Microwave Brightness Temperatures over Snow-Covered Terrain in High Mountain Asia: What are the Sensitivities and Potential Pitfalls of Machine Learning?

JAWAIRIA A. AHMAD[1] AND BARTON A. FORMAN[1]

ABSTRACT

Snow and ice melt from high mountain Asia (HMA) provides freshwater on which over 136 million people depend for their basic needs. Despite its importance in life sustenance, there is still considerable uncertainty regarding the spatial and temporal variation of snow in HMA. In this study, Noah MP is used within the NASA Land Information System (LIS) framework to model the hydrologic cycle over the Indus basin. The capability of support vector machines (SVM), a machine learning technique, to predict passive microwave brightness temperatures (T_b) as a function of LIS modeled geophysical states is explored through a sensitivity analysis. Passive microwave brightness temperatures, as measured by the Advanced Microwave Scanning Radiometer - Earth Observing System (AMSR-E) over snow-covered terrain in the Indus basin, are used as training targets during the SVM training process. Normalized sensitivity coefficients (NSC) are computed to assess the sensitivity of a well-trained SVM to each LIS modeled state variable. Sensitivity analysis results conform with the known first-order physics, i.e., geophysical states that are directly related to physical temperature have relatively higher NSC magnitudes. Irrational NSC signs are observed in some cases that are explored in detail. General adherence of the SVM approach to the first-order physics bodes well for its potential use in LIS as the observation operator within a brightness temperature data assimilation framework aimed at improving snow estimates. However, pitfalls do exist that serve to highlight the limitations of this particular machine learning algorithm, but these pitfalls can be avoided with proper consideration of the first-order physics.

[1] University of Maryland, College Park, MD, USA

Poster Session

New Cloud Mask Algorithm over Snow/Ice-Covered Areas based on Machine Learning Techniques and Comprehensive Radiative Transfer Simulations

NAN CHEN[1], WEI LI[1], CHARLES GATEBE[2], TOMONORI TANIKAWA[3], MASAHIRO HORI[4], TERUO AOKI[3,5], RIGEN SHIMADA[3,4], AND K. STAMNES[1]

ABSTRACT

Cloud detection and screening constitute critically important first steps required to derive many satellite data products. Traditional threshold-based cloud mask algorithms require a complicated design process and fine tuning for each sensor, and they have difficulties over areas partially covered with snow/ice. Exploiting advances in machine learning techniques and radiative transfer modeling of coupled environmental systems, we have developed a new cloud mask algorithm based on a neural network classifier driven by extensive radiative transfer simulations. Statistical validation results obtained by using collocated CALIOP and MODIS data show that its performance is consistent over different ecosystems and significantly better than the MODIS Cloud Mask (MOD35 C6) during the winter seasons over snow-covered areas in the mid-latitudes. Simulations using a reduced number of satellite channels also show satisfactory results, indicating its flexibility to be configured for different sensors. Compared to threshold-based methods and previous machine-learning approaches, this new cloud mask (i) does not rely on thresholds, (ii) needs fewer satellite channels, (iii) has superior performance during winter seasons in mid-latitude areas, and (iv) can easily be applied to different sensors.

[1] Stevens Institute of Technology, Hoboken, NJ, USA
[2] NASA Goddard Space Flight Center, Greenbelt, MD, USA
[3] Meteorological Research Institute, Tsukuba, Japan
[4] Earth Observation Research Center, Japan Aerospace Exploration Agency, Tsukuba, Japan
[5] Okayama University, Okayama, Japan

What is Winter? A Socio-Ecological Reckoning

ALEXANDRA R. CONTOSTA[1], NORA CASSON[2], SARAH J. NELSON[3], AND SARAH GARLICK[4]

ABSTRACT

In temperate, seasonally snow-covered ecosystems globally, winter is increasingly recognized as a critical period for regulating ecological, biogeochemical and socio-economic processes, both during the winter season and throughout the year. In these areas, winters are characterized by below freezing temperatures and sustained snowpacks, both of which are disappearing. Traditional seasonal definitions based on astronomical or meteorological reckonings are out of sync with this socio-ecological experience of weather and do not capture how the timing of winter could shift in a context of changing climate. Here we propose a new reckoning of winter that can vary in space and time: we define a **socio-ecological winter** as a period of sustained temperatures below freezing and snow accumulation that together regulate ecological and biogeochemical processes and the services they provide to society, both during winter and annually. To explore this new reckoning, we examined 100 years of meteorological data across seasonally snow-covered areas of eastern North America. Trend analysis demonstrates that winters have become shorter across the region, at a rate of up to two days per decade. This is consistently due to an earlier onset of spring, although at many sites there is also a significant trend towards a later end of fall. Within winters, conditions have become more variable, with significant increases in the number of days above freezing temperatures, that winter rain occurs, and that bare ground exists instead of snow cover. Together, these indicators show a trend toward losing the conditions that define winter across our region.

[1] University of New Hampshire, Durham, NH, USA
[2] University of Winnipeg, Winnipeg, MB, Canada
[3] University of Maine, Orono, ME, USA
[4] Hubbard Brook Research Foundation, Woodstock, VT, USA

Eastern Snow Conference Meeting Locations

MILES ECCLESTONE[1]

ABSTRACT

Since its founding in 1940 the Eastern Snow Conference has been held at many locations across Canada and the United States, many of which are shown on this map.

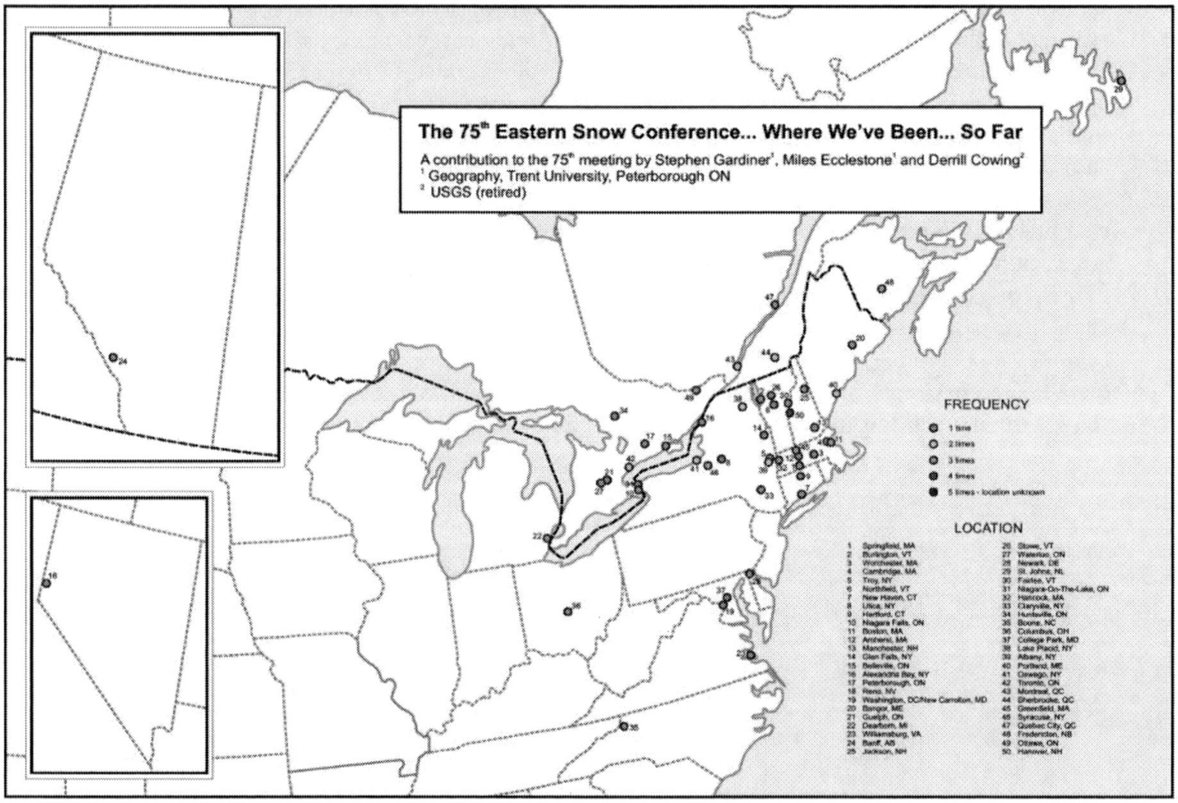

---

[1] Trent University, Peterborough, ON, Canada

Assessment of Advanced Technology Microwave Sounder (ATMS) Snow Products

CHRISTOPHER GRASSOTI[1] AND SEAN R. HELFRICH[1]

ABSTRACT

The Advanced Technology Microwave Sounder (ATMS) is a cross-track scanning instrument onboard the Suomi National Polar-orbiting Partnership (S-NPP) satellite. ATMS has 22 channels ranging from 23 GHz through 183 GHz. Products generated from ATMS retrievals use the Microwave Integrated Retrieval System (MIRS) algorithm, a physically-based retrieval (1DVAR) capable of retrieving atmospheric and surface parameters concurrently. While not designed to observe surface snow, MIRS does provide estimations of snow water equivalent (SWE) and snow grain size (SGS). A recent examination has explored algorithm adjustments for forest fraction, a common source of error in passive microwave snow retrievals. Past efforts to correct for forest fraction applied a snow climatology. However, conditions in the forested ephemeral snow zones tend to remain underestimated in dense forests when the values exceed climatology. This poster demonstrates the potential for forest surface fraction and forest type to improve SWE estimates from ATMS. ATMS SWE was compared to SWE estimated from Advanced Microwave Sounding Unit (AMSU) from NOAA 20, Japan Aerospace Exploration Agency (JAXA) Advanced Microwave Scanning Radiometer – 2 (AMSR 2), and other derived products. Initial results are promising and suggest that implementation should compensate for poor estimates in heavily forested areas, particularly when the snow is wet.

[1] NOAA / NESDID / STAR, College Park, MD, USA

Bicontinuous Dense Media Radiative Transfer (DMRT) Model for Applications to Snow Parameters Retrievals in Satellite Microwave Remote Sensing and Data Assimilation

WEIHUI GU[1], JIYUE ZHU[1], SHURUN TAN[1], AND LEUNG TSANG[1]

ABSTRACT

Snow is a vital water source for natural environment and human society. Knowledge of snow distribution is of great importance for water resource management and climate change prediction. Using microwave remote sensing technique to monitor snowpack has drawn attention in the snow community. A physical model that can accurately predict microwave signals such as backscatter and brightness temperature is useful for interpreting the microwave signatures. However, the physical models can be computational intensive requiring numerical solutions of Maxwell equations. To use the microwave model for real time retrieval of snow parameters and for coupling with land surface models and hydrology models for data assimilation, the microwave output need to be obtained in real time.

Previously we have posted the QCA-DMRT model that represents snow using sticky spheres. In this paper, we describe the on-line microwave physical model of bicontinuous medium that has been recently posted on *http://web.eecs.umich.edu/~leutsang/Computer%20Codes%20and%20 Simulations.html*. A merit of the bicontinuous medium model is that the computer-generated microstructure resembles snow. We describe the procedures to run the Bicontinuous DMRT model (Bic/DMRT) so that accurate microwave signal outputs can be obtained. The Bic/DMRT open source code is a toolkit for modeling backscatters and brightness temperatures of multi-layer snowpack over rough soil/sea ice surface. A lookup table (LUT) containing phase matrix, extinction coefficient, effective permittivity of snow is built. With the pre-built LUT, the simulation time is only for solving DMRT equations. Outputs in the LUT can be transformed and used in other radiative transfer solvers as well. The input for Bic/DMRT is a script describing the physical information of multi-layer snowpack, including snow depth, volume fraction, temperature, snow microstructure parameters $\langle\zeta\rangle$ and b for each layer. $\langle\zeta\rangle$ with unit of m^{-1} is inversely proportional to the mean grain size of snow. The dimensionless parameter b is negatively correlated to the standard deviation of ζ. It describes the aggregation of snow particles and affects the tail of the correlation function. Two routines corresponding to passive and active DMRT are used to solve for brightness temperatures and backscatters, respectively. The simulation time is linearly proportional to the number of snow layer. By reducing the number of quadrature angles in computing the coupled DMRT equations while maintaining the accuracy of results, the simulation time for a single layer is within a second. Higher efficiency can be achieved by further optimization. The posted code is written in Matlab and can be further modified for broader applications.

[1] University of Michigan, Ann Arbor, MI, USA

Observations of Snow Particle Characteristics during Snow Events in the Southern Appalachian Mountains

HEATHER GUY[1], L. BAKER PERRY[1], AND SANDRA E. YUTER[2]

ABSTRACT

In the southern Appalachian Mountains, the timing and intensity of snow events is difficult to predict due to the complicated orography and the wide range of synoptic setups that can result in snowfall. Unexpected snowfall totals can still have widespread economic, social and environmental consequences in this region. More generally, the large uncertainties surrounding cloud microphysical properties and their relationship with cloud structure and snowfall totals remains one of the key limitations in weather and climate models. This study presents case studies of snow events that occurred between November 2017 and April 2018 in Boone, NC (1,016 m asl). Digital photographs of snow particle characteristics taken by a Multi-Angle Snowflake Camera (MASC) are assessed alongside data from a vertically pointing Micro Rain Radar (MRR) to explore how vertical structure impacts ice crystal formation and riming during these events. Manual snowfall measurements, data from a network of meteorological stations in the Southern Appalachian region, and synoptic charts allow the assessment of the synoptic conditions associated with each event. Additionally, hourly snow samples were collected during three storms and analyzed for their isotopic ($\delta^{18}O/\delta D$) composition. The development of the isotopic composition of the snow throughout these storms provides additional information about the conditions during snow crystal formation and into the relationship between cloud microphysics and the isotopic composition of precipitation. The goal of this study is to present a detailed analysis of the cloud microphysical properties associated with the different synoptic patterns during each storm, the resulting snowfall totals, and the impact of each event.

[1] Appalachian State University, Boone, NC, USA
[2] North Carolina State University, Raleigh, NC, USA

Passive Microwave Remote Sensing of Colorado Watersheds using Calibrated, Enhanced-Resolution Brightness Temperatures (CETB) for Estimation of Snowmelt Timing - CLPX and SnowEx

MITCHELL JOHNSON[1], JOAN RAMAGE[2], TARA J. TROY[1], AND MARY J. BRODZIK[3]

ABSTRACT

Understanding the timing of seasonal melt onset is critical for water resources management in snow-dominated watersheds. Passive microwave remote sensing has been used to detect melt-freeze events with the Special Sensor Microwave Imager (SSM/I) and the Advanced Microwave Scanning Radiometer - Earth Observing System (AMSR-E) with global coverage. This study investigates the newly available passive microwave calibrated, enhanced-resolution brightness temperature datasets (CETB) produced at the National Snow and Ice Data Center to estimate melt timing at higher spatial resolution (~3 km). CETB datasets generated from SSM/I and AMSR-E are used to characterize snowmelt timing in mountainous Colorado basins that were part of NASA's Cold Land Processes Field Experiment (CLPX 2002-2003) and SnowEx 2017 campaigns. We employ existing algorithms using the diurnal amplitude variation (DAV) and cross-polarized gradient ratio (XPGR) methods at 36 GHz and 18 GHz to detect seasonal melt onset date and early season melt occurrences. Comparisons of algorithm results with ground observations determine the optimal melt onset algorithm thresholds for the newly processed data. We show that the higher-resolution datasets yield an improvement in snowmelt detection in landscapes with heterogeneous topography and land cover. This work provides insight into the performance of higher-resolution reprocessed CETB data for snowmelt analysis and will enable hydrologists to better analyze the characteristics and implications of snow melt in mountainous watersheds.

[1] Department of Civil and Environmental Engineering, Lehigh University, Bethlehem, PA, USA
[2] Department of Earth and Environmental Science, Lehigh University, Bethlehem, PA, USA
[3] National Snow and Ice Data Center, Boulder, CO, USA

Snow Ensemble Uncertainty Project (SEUP): Quantification of Snow Water Equivalent Uncertainty across North America via Ensemble-Based Land Surface Modeling

RHAE SUNG KIM[1], SUJAY KUMAR[2], CARRIE VUYOVICH[3], PAUL HOUSER[4], MICHAEL DURAND[5], GLEN LISTON[6], JESSICA LUNDQUIST[7], EDWARD KIM[2], ANA BARROS[8], CHRIS DERKSEN[9], BARTON A. FORMAN[10], CAMILLE GARNAUD[11], AND MELODY SANDELLS[12]

ABSTRACT

Well-characterized snow water equivalent (SWE) uncertainty is critical in seasonal to decadal climate prediction systems as well as for planning future snow satellite missions. In this study, an ensemble-based land surface modeling approach is applied to characterize the sources and regions of high model uncertainty across North America. Four different land surface models of varying complexity are driven using three different forcing datasets. Model simulations are conducted across North American over multiple winter snow seasons at a 5-km spatial resolution using the NASA Land Information System (LIS). Additionally, a distributed snow-evolution modeling system (SnowModel) using the output forcing data from LIS is run independently. The relatively coarse resolution meteorological inputs are then spatially downscaled using a mix of lapse-rate, slope-aspect, and climatology-based approaches. The Hydrological Modeling and Analysis Platform (HyMAP) in LIS is the used to derive estimates of streamflow. In this study, we characterize the dominant factors that govern the spatial variability and seasonality of SWE uncertainty. The contributing role of various landscapes and precipitation regimes on the modeled SWE estimates will also be quantified. The results of the study are expected to guide the selection of sites for future snow field campaigns as well as provide useful guidance towards the planning of future satellite missions.

[1] Earth System Science Interdisciplinary Center, College Park, MD, USA
[2] NASA Goddard Space Flight Center, Greenbelt, MD, USA
[3] U.S. Army Engineer Research and Development Center, Hanover, NH, USA
[4] George Mason University, Fairfax, VA, USA
[5] The Ohio State University, Columbus, OH, USA
[6] Colorado State University, Fort Collins, CO, USA
[7] University of Washington, Seattle, WA, USA
[8] Duke University, Durham, NC, USA
[9] Climate Research Division, Environment and Climate Change Canada, Toronto, ON, Canada
[10] University of Maryland, College Park, MD, USA
[11] Meteorological Research Division, Environment and Climate Change Canada, Dorval, QC, Canada
[12] CORES Science and Engineering Limited, Burnopfield, UK

Feasibility of a Microwave Brightness Temperature Data Assimilation Framework using the NASA Land Information System and a Well-Trained Support Vector Machine to Improve Snow Water Equivalent Estimates over High Mountain Asia

YONGHWAN KWON[1], BARTON FORMAN[1], YEOSANG YOON[2], AND SUJAY V. KUMAR[2]

ABSTRACT

This study explores the potential for using a passive microwave brightness temperature data assimilation (DA) framework to enhance the characterization of spatiotemporal variability of snow water equivalent (SWE) across High Mountain Asia. The DA framework includes the NASA Land Information System in conjunction with a well-trained support vector machine (SVM) that acts as the observation operator. The Noah Land Surface Model with multi-parameterization options (Noah-MP) is used as the prognostic land surface model. In view of the lack of available measurements for validation in the study area, a synthetic twin experiment is adopted. The DA framework for the synthetic experiment is composed of two phases: (1) SVM training and generation of synthetic observations, and (2) assimilation using an ensemble Kalman filter. Each "twin" uses the same Noah-MP model but is forced by a different reanalysis product for the periods 01 Sep 2002 to 01 Sep 2011. Synthetic observations of spectral brightness temperature difference, i.e., ΔT_B between 10.65- and 36.5-GHz horizontal polarization (H), between 10.65- and 36.5-GHz vertical polarization (V), between 18.7- and 36.5-GHz H, and between 18.7- and 36.5-GHz V, produced during phase 1 are then assimilated into Noah-MP using an ensemble Kalman filter during phase 2. A series of experiments are conducted to address three research questions: (1) How well are the SVM-based ΔT_B predictions correlated with the simulated SWE? (2) Can the assimilation of multivariate observations improve DA performance compared to univariate assimilation? (3) How does the existence of glaciers influence the behavior of the SVM-based ΔT_B predictions and the corresponding DA performance?

[1] University of Maryland, College Park, MD, USA
[2] NASA Goddard Space Flight Center, Greenbelt, MD, USA

Improving the Understanding and Uncertainty of Snow Radiative Transfer Modeling using Snowpack Information of Varying Complexity

THEODORE LETCHER[1] AND CARRIE VUYOVICH[1]

ABSTRACT

Spatial characterization of snow water equivalent using spaceborne passive microwave remote sensing platforms is critical for understanding the hydrological, meteorological, and mobility environments in regions where quality snow observations are lacking. However, using passive microwave data to determine the amount of snow liquid on the ground remains a significant challenge, especially over mountainous regions with deep snow packs and dense forests. Recently, snow radiative transfer models have been developed to better understand how microwave radiation is affected by various snowpack properties, and to facilitate data assimilation and land model initialization.

One knowledge gap that exists in this area is that the most widely used operational land surface models only provide bulk snow pack information, such as snow depth and snow water equivalent, rather than detailed information about the snowpack stratigraphy and snow grain size. This study attempts to address this knowledge gap by simulating snow brightness temperatures using the Dense Media Radiative Transfer Multi-Layer Model (DMRTML) using various degrees of complexity with respect to the snowpack information used to drive the DMRTML.

The snow and radiometer data used to force and evaluate the DMRTML were collected during the CrustEX and SnowEx field campaigns in central New Hampshire in February 2018 and in Western Colorado in February 2017. Ways to incorporate more complex snow pack information critical to radiative transfer into common land surface models are discussed.

This work shines a light on what level of complexity is required in a snow model and what specific snow characteristics are critical to realistically simulating snow brightness temperatures towards the assimilation of satellite microwave data.

[1] U.S. Army Corps of Engineers Cold Regions Research and Engineering Laboratory, Hanover, NH, USA

Status of the GOES-R Fractional Snow Cover Product

YINGHUI LIU[1], JEFFREY R. KEY[2], AND AARON LETTERLY[1]

ABSTRACT

The global coverage of polar-orbiting satellites and the high temporal frequency of geostationary satellites allow for the robust characterization of the Earth's snow cover. Unlike binary snow maps that provide only a snow/no-snow labelling, satellite-derived Fractional Snow Cover (FSC) products provide estimates of the area fraction of a sensor field-of-view covered by snow. For U.S. coverage, snow cover products based on a single reflective channel of the Geostationary Operational Environmental Satellite (GOES) Imager have been available for many years. Now, with the enhanced spectral sampling and higher spatial resolution that the GOES-R Advanced Baseline Imager (ABI) provides, we can expect improvements in the level of detail and accuracy of geostationary snow products.

The algorithm used for GOES-R FSC is based on the Moderate Resolution Imaging Spectroradiometer (MODIS) "MODSCAG" method developed by T. Painter. For ABI it is known as the GOES-R Snow Cover And Grain size (GOESRSCAG) algorithm. The product includes not only fractional snow cover but also snow grain size. It employs an optimized spectral mixture analysis using atmospherically-corrected, surface spectral reflectance in two visible, five near-visible, and one thermal wavelengths. Though the GOES-R ABI surface reflectance product is still under development, the GOESRSCAG product that is routinely generated using reflectances at the top of atmosphere is giving promising results. In this presentation, we will provide more detail on the current status of the GOES-R FSC and future calibration/validation plans. NOAA's goal to select a single "enterprise" snow algorithm for use with multiple satellite instruments will also be discussed.

[1] University of Wisconsin, Madison, WI, USA
[2] National Oceanic and Atmospheric Administration, Madison, WI, USA

Inversion of Snow Depth from UAVSAR L-band PolSAR Data

SURENDAR MANICKAM[1], AVIK BHATTACHARYA[2], AND MATTIAS BRAUN[1]

ABSTRACT

In this work, UAVSAR L- band Polarimetric Synthetic Aperture Radar (PolSAR) data along with the SnowEx 2017 field campaign data are utilized for the development and validation of a new snow depth estimation algorithm. The UAVSAR data acquired on 22 February 2017 over Grand Mesa are used. The SnowEx raw penetration force profiles measured at snow pits at Grand Mesa, Colorado using the SnowMicroPen (SMP), a digital snow penetrometer on the same date with the UAVSAR data are used for the analyzes and validation of the snow depth estimations.

Different PolSAR parameters are analyzed with the corresponding snow depth measurements. Out of those parameters, it is found that dominant scattering type phase ($\Phi s1$) and helicity (T1) from Touzi incoherent polarimetric decomposition method [1] are providing useful information about the snow depth. These parameters are thoroughly investigated. Finally, based on the investigations, a new generalized polarimetric parameter for the snowpack is developed. This generalized parameter is inverted as a snow depth parameter. The investigation shows that the estimated snow depth from the proposed approach is having a high correlation with the measured snow depth values. However, other snowpack parameters collected in the field along with the different snow covered ground surfaces to be analyzed and taken into the account for the better estimation of the snow depth.

REFERENCES

Touzi R. 2007. Target scattering decomposition in terms of roll-invariant target parameters. IEEE Transactions on Geoscience and Remote Sensing, 45(1): 73-84.

[1] Friedrich-Alexander-Universität Erlangen-Nürnberg, Germany
[2] Indian Institute of Technology Bombay, Mumbai, India

Dual-pol Passive Coherent Measurement of Snow-on-Ice near Grazing with WiBAR

SEYEDMOHAMMAD MOUSAVI[1] AND ROGER DE ROO[1]

ABSTRACT

WideBand Autocorrelation Radiometry (WiBAR) is a passive microwave technique to measure the thickness of low-loss layers such as snow pack and lake ice pack. For wavelengths sufficiently long that the surfaces appear smooth and the scattering is negligible, the microwave brightness emitted from the first lossy material (the ground, or water) has at least two paths of different lengths to the microwave remote sensor. One path is directly upward through the low loss layers to the antenna; another path involves the reflection of the upward traveling brightness by the layer upper surface, then again by the layer lower surface, before traveling to the antenna. The noise in the second path is an attenuated and delayed copy of that in the first, and the delay, revealed as a non-zero local maximum in the temporal autocorrelation function (ACF) of the received waveform, is the observable signal.

When multi-layer structures exist, such as a snow pack on lake ice, multiple paths exist, which may confound the interpretation of the ACF. Dual pol measurements can be used to resolve the ambiguities. Due to its significantly higher reflection coefficients, H-pol often has observable delays regardless of incidence angle (nadir to at least 85deg), while V-pol often has no observable delays. However, near grazing, the V-pol reflection coefficients become sufficiently large for some of the interfaces in a snow over ice scene, while the H-pol reflection coefficients remain sufficiently large for most interfaces. As a result, the two polarizations provide complimentary information about the scene, and the snow depth, which could be too thin to observe directly in the ACF, can be retrieved from dual polarization observations at the same grazing angle. Examples from measurements on South Sturgeon Lake, MN, USA on 2018 Mar 07 illustrate these principles.

[1] University of Michigan, Ann Arbor, MI, USA

Snow Water Equivalent Synthetic Aperture Radar and Radiometer (SWESARR)

BATUHAN OSMANOGLU[1], RAFAEL RINCON[1], QUENTON BONDS[1], PAUL RACETTE[1], AND LUDOVIC BRUCKER[1,2]

ABSTRACT

Snow water equivalent (SWE) is a quantity of paramount importance in the hydrologic cycle since it is directly related to the amount of freshwater stored as snow that becomes available each spring for drinking water, farming, and electricity production. SWE is a highly challenging quantity to estimate using remote sensing techniques, due to its variation based on many geophysical variables (snow depth, grain size, density, soil moisture, etc.). While snow depth can be estimated from surface height differencing of snow off and snow on conditions using lidar observations, this technique is insufficient to estimate SWE.

SWESARR is a three-band radar and three-band radiometer instrument developed specifically to improve our understanding of snow water equivalent estimation using microwave data. SWESARR employs three radar bands (9.6, 13.6, and 17.2 GHz) and three radiometer bands (10.6, 18.7 and 36.5 GHz) that penetrate the snow and allow us to constrain the geophysical parameters to accurately estimate SWE. In addition, radar data is collected in dual polarization (VV, VH) while the radiometer makes single polarization (H) observations. The combination of all these microwave measurements will provide an important data set to improve SWE retrieval algorithms.

SWESARR radiometer has been developed since 2010, and the radar is in active development. During the SnowEx 2017 campaign SWESARR radiometer flew two science flights over the Grand Mesa, CO on February 9 and 10. In this presentation, we will describe SWESARR, explain its capabilities and present data that was collected by the radiometer.

[1] NASA Goddard Space Flight Center, Greenbelt, MD, USA
[2] Universities Space Research Association, Columbia, MD, USA

Towards the Assimilation of C-band Synthetic Aperture Radar (SAR) Backscatter Observations over Snow-covered Terrain

JONGMIN PARK[1] AND BARTON A. FORMAN[1]

ABSTRACT

Estimating snow mass using space-based synthetic aperture radar (SAR) as part of a data assimilation framework holds many challenges. However, the all-weather capability of C-band SAR (as opposed to optical or thermal channels) coupled with the fine spatial resolution of the active microwave system (as opposed to passive microwave systems) make space-based SAR an attractive option.

This study explores the relationship between terrestrial snow depth, snow water equivalent, and Sentinel-1 C-band backscatter. Sentinel-1 is a constellation of two satellites with a 180-degree phase difference (Sentinel-1A and Sentinel-1B) that is operated by the European Space Agency. Among the different acquisition modes and product types, the Sentinel-1 interferometric wide swath (IW) ground range detected (GRD) dataset was used in this research. The observations were preprocessed through a series of steps accounting for sensor orbit, thermal and speckle noise, radiometric calibration, and terrain correction using ESA's Sentinel Application Platform (SNAP) software package. Preprocessed backscatter coefficients were then compared against snow depth measurements from the Global Surface Summary of the Day (GSOD) and Snow Telemetry (SNOTEL) stations within our study area near Grand Mesa, Colorado. Grand Mesa was selected in this study to leverage existing measurement networks as well as to better harness on-going NASA Snow Experiment (SnowEx) Campaign activities in and around the region. Comparison of Sentinel-1 backscatters with ground-based observations is an essential first step that will help characterize the uncertainty (and error) of Sentinel-1 observations, which is an important precursor for backscatter assimilation within an ensemble-based, multi-variate data assimilation framework.

[1] University of Maryland, College Park, MD, USA

The Nexus of an Alpine Glacier Watershed, Climate Change and Human Activity: Nooksack River, Washington

MAURI S. PELTO[1]

ABSTRACT

Alpine glacier watersheds are changing significantly, this is not happening at a "glacial" pace, and the regional impacts are profound. The rapid decline of alpine glacier area and volume is a consistent trend from mountain range to mountain range, emphasizing that though regional climate and glacier response differ, global changes are driving the response. The trend of glacier loss is expected to increase with continued anthropogenic warming further altering the timing and magnitude of glacier runoff. Mountain glaciers are important as water resources for agriculture, hydropower, aquatic life and municipal water supply, melting fastest in the summer when precipitation is lowest and water demand from society is largest. The specific cascade of impacts downstream varies between glacier fed watersheds. Here we use the Nooksack River, WA to illustrate the changes.

[1] Nichols College, Dudley, MA, USA

Adapting Model Representation of Liquid Water Percolation in Maritime Environments

JUSTIN M. PFLUG[1], GLEN E. LISTON[2], AND JESSICA D. LUNDQUIST[1]

ABSTRACT

Snowpack liquid water percolation is a dominant process throughout the ablation and accumulation seasons. However, past work has shown that commonly-modeled snowpack properties like hydraulic conductivity, liquid water content, and ice content can determine infiltration from liquid sources like snowmelt, rain, and canopy drip. Errors in thermal inertia, precipitation partitioning, and preferential flow can make percolation notoriously difficult to model. This problem is compounded in maritime regions known for accumulation-season melt, frequent canopy drip, and rain-on-snow events. In this study, we adapted SnowModel to include a gravity-drainage flow representation of liquid water percolation. SnowModel was chosen as this model was sufficiently able to resolve snowpack properties for both single and multilayer schemes without changing the underlying model processes or order of operations. The model adjustments were evaluated in both maritime and continental climates over multiple years to evaluate transferability and applications to global modeling. Sensitivities to layering schemes, precipitation partitioning, and numerical solvers were investigated throughout. Results indicate that the best performance resulted from percolation derived from gravity drainage flow with multiple layers. Bulk snowpack gravity drainage also provided favorable results for models unable to resolve multiple layers.

[1] University of Washington, Seattle, WA, USA
[2] Colorado State University, Fort Collins, CO, USA

Enhanced 30-year Global Snow and Ice Dataset and Climatology derived from Combined Satellite Observations in the Visible/Infrared and Microwave Spectral Bands

PETER ROMANOV[1,2]

ABSTRACT

Since 2006 NOAA has been using Global Multisensor Automated Snow and Ice Mapping System (GMASI) for operational monitoring of the Earth's cryosphere. The GMASI system implements an automated algorithm which processes observations in the visible and infrared bands from the AVHRR sensor onboard NOAA and METOP satellites and observations in the microwave from SSMI and SSMIS sensors onboard DMSP satellites. The primary output of the system presents a daily spatially-continuous (gap-free) maps of snow and ice cover at the nominal spatial resolution of about 4 km. In this study we have employed the GMASI algorithm to reprocess historical satellite data back to 1987 and to generate a high-resolution 30-year long daily dataset of global snow and ice maps. Daily maps have been further used to produce the snow and ice cover climatologies including their monthly mean frequency of occurrence as well as multiyear trends in the snow and ice extent.

In the presentation we provide details on the data processing algorithm as well as on the calibration adjustment of the visible/infrared and microwave sensors data to achieve consistency of retrievals across the whole 30-year long time period. Estimates of the snow and ice extent and corresponding long-term trends derived from the GMASI-reprocessed data are compared with the available coarse resolution snow cover climatology based on NOAA interactive snow and ice charts.

[1] City College of City University of New York, New York, NY, USA
[2] Center for Satellite Applications and Research, NOAA/NESDIS, College Park, MD, USA

SnowEx 2017 In-situ Passive Microwave Measurements: Analysis of Wet Snow Microwave Emission

ALEXANDRE ROY[1,2], ALEXANDRE LANGLOIS[1,2], CAROLINE DOLANT[1,2], LUDOVIC BRUCKER[3,4], AND ALAIN ROYER[1,2]

ABSTRACT

Snow is a key hydrological element, acting as an important freshwater reservoir for human consumption and hydroelectricity production. Large uncertainties remain, however in regard to: 1) the amount and distribution of water stored in seasonal snow cover (snow mass); 2) the effect of warming temperatures on climatological cooling and heating patterns, including temporal lags and spatial teleconnections (snow extent); 3) the initialization of snow cover for operational numerical weather predictions, hydrological simulations, and seasonal forecasts; and 4) the parameterization of snow physics in global circulation models (GCMs) for climate projections. All of these areas would benefit from improved snow products derived from remote sensing. With these potential remote sensing tools available, a multi-scale remote sensing and modeling approach becomes essential, and in-situ validation of airborne and satellite measurements must be conducted over a wide range of snow, vegetation, and climatic conditions.

During the NASA SnowEx 2017 campaign, four surface-based radiometers (SBR) operating at 89, 37, 19, and 10.67 GHz [1] were deployed on Grand Mesa, Colorado from February 14th to February 18th. This paper presents an overview of the microwave radiometer measurements recorded in about 50 locations. In addition, due the warm condition that week, the effect of liquid water in snow on multi-frequency passive microwave signal is investigated.

[1] Université de Sherbrooke, Sherbrooke, QC, Canada
[2] Centre d'Étude de la Neige
[3] NASA Goddard Space Flight Center, Greenbelt, MD, USA
[4] Universities Space Research Association, GESTAR, Columbia, MD, USA

Exploration into the Potential Linkage between Local Fluctuations in Passive Microwave Snow Water Equivalent (SWE) Retrieval and Various Characteristics of a Rain-on-Snow (ROS) Event

E. MEGHAN RYAN[1], LUDOVIC BRUCKER[2,3], AND BARTON A. FORMAN[1]

ABSTRACT

Wintertime rain-on-snow (ROS) events can impact snow stratigraphy via generation of wet snow and ice crust(s) either on or within the snowpack. Considering the assumptions of most passive microwave-based snow water equivalent (SWE) retrievals, which include a dry and homogenous snowpack, ROS events could significantly impact the accuracy of said SWE retrievals. Relatively little is known about the location and occurrence of most high-latitude ROS events. Consequently, in the absence of ground-based observations, it is difficult to discern if satellite-based SWE retrievals are doing an adequate job characterizing ROS-impacted snow packs. This study investigated potential linkages between local SWE fluctuations and the particular characteristics of an ROS event to further explore the efficacy of SWE retrievals at locations where an ROS event may have occurred.

ROS event characteristics explored include daily-averaged snow depth, precipitation, and near-surface air temperature. ROS events were also examined in terms of their timing in the winter season and spatial location. ROS events studied were detected using spectral and temporal criteria on variables calculated from the Advanced Microwave Scanning Radiometer – Earth Observing System (AMSR-E) observed brightness temperatures. Using the NSIDC AMSR-E/Aqua L3 Global SWE product, daily changes in SWE before, during, and after detected ROS events were collected and subsequently compared to those of ROS events with similar characteristics. This research has the potential to inform passive microwave SWE retrievals about snow pack conditions that violate the inherent retrieval assumptions, and hence, may require additional modification or flagging at a specific location in space and time.

[1] University of Maryland, College Park, MD, USA
[2] NASA Goddard Space Flight Center, Greenbelt, MD, USA
[3] Universities Space Research Association, Columbia, MD, USA

Detection of Snowmelt Signals for Improving Snowmelt Flood Forecasts in the Red River Basin of the North

RONNY SCHROEDER[1], SIMON KRAATZ[1], JENNIFER M. JACOBS[1], BRIAN CONNELLY[2], AND MICHAEL M. DEWEESE[2]

ABSTRACT

NOAA's North Central River Forecast Center and the University of New Hampshire (UNH) have partnered to improve snowmelt flood predictions for the Red River basin of the North (RRB). UNH delivers daily satellite passive microwave SWE observations in a format that is readily ingested into their operational computing environment. While the satellite SWE provides valuable information in this data sparse region, their utility in an operational flood forecasting context is hampered because of wet snow contamination. This presentation describes a new QC protocol that detects wet snow signal contamination in the RRB. Here, we demonstrate the signal detection scheme's identification of wet snow events using 3 years (2015-2018) of daily, satellite Special Sensor Microwave Imager/Sounder (SSMIS)) SWE time series in the RRB. The technique applies a dynamic thresholding algorithm to identify abrupt decreases in the SWE time series associated with wet snow. The method accounts for sudden increases as well as for less pronounced variations in SWE and is less likely to identify such events as snow melt signals in the SWE time series. We evaluate the method's performance by comparing the time series of snow melt events with independent melt signals from coincident SMAP L-band radiometer data using a threshold on polarization difference. We examine temporal consistency between the two methods and discuss the effect of differences in sampling frequency with respect to sampling depth and the identification of liquid water in the snowpack.

[1] University of New Hampshire, Durham, NH, USA
[2] NOAA/NWS/North Central River Forecast Center, Chanhassen, MN, USA

Avalanche in Eastern Canada: A Provincial Responsibility

JERRY TOUPIN[1]

ABSTRACT

Eastern Canada remains one of the snowiest places in the world. While most of it gets over 2 meters of snow every winter, parts of Québec, Labrador and Newfoundland receive over 3 meters of snow between October and April. Every year the risk of avalanche remains high. An avalanche can fall at a speed of more than 250 kilometers per hour and carry over 100,000 tons of snow and other debris. These massive snow slides can then be very dangerous and even deadly. Nain, a small mostly Inuit community located in Labrador (56° 32' N, 61° 41' W) and one of the most northernmost permanent settlements of the province of Newfoundland and Labrador, is the earliest recorded avalanche for Canada during the winter of 1781-1782 claiming 22 lives.

New technology is being developed to prevent avalanche and to save lives. Governments and stakeholders are regrouping to better protect people and infrastructure from damage caused by avalanche.

Keywords: Eastern Canada, Snow, Avalanche, Provincial ministries and agencies, History.

INTRODUCTION

Although not as common and frequent (Toupin, 2018) as in Western Canada (that includes: The Canadian Rockies in Alberta, British Columbia (Revelstoke gets about 1200 cm of snow every year), Yukon and the Northwest Territories), avalanche is nevertheless a problem (Hétu *et al.*, 2008) in Eastern Canada (mostly in Quebec with about 70 deaths and in Newfoundland-Labrador with a total of deaths of 75 since the beginning of the 19th century) for infrastructures and human activities and in Nova Scotia (Cape Breton Island).

Unfortunately, more and more problems are expected in the near future as a greater number of people are involved in backcountry activities. The role of every provincial government and other stakeholders is crucial to save lives during the cold season.

ONTARIO AND QUÉBEC

Rare in Ontario, although at least one deadly avalanche occurred in Toronto according to the Canadian Encyclopedia, it is the Ministry of the Environment, Conservation and Parks as well as the Ministry of Natural Resources and Forestry that monitor snow and ice conditions among others and protect people from potential deadly avalanche in that province.

In the province of Québec, since 1999 Avalanche Québec a non-profitable organization was founded to monitor snow condition and forecasting avalanche for Eastern Québec (Appalachian

[1] Retired, University of Alberta, Edmonton, AB, Canada

region, Chic-Choc). Training, information and awareness are also available for people. Departments and agencies such as Municipal Affairs, Land Use, Energy and Natural Resources, Sustainable Durability, Environment and Global Warming, Public Safety, Forestry, Parks and Fauna are all working hands in hands to better prevent and monitor avalanche.

Table 1. Number of fatalities (examples) in Eastern Canada due to avalanche.

Year	Number	Example
1791-82	22	Canada's earliest recorded avalanche, Nain, Labrador
1877	6	Baie verte, Newfoundland
1912	5	Tilt Cove, Newfoundland
1935	2	Okak, Labrador
1999	9	Kangiqsualujjuaq, Québec
2000	2	Monts Chic-Choc, Québec
2007	1	Blue Mountain, Newfoundland

NEWFOUNDLAND, LABRADOR AND NOVA SCOTIA

The province of Newfoundland and Labrador is the most at risk for avalanche (Liverman, 2018) in Eastern Canada as close to 80 deaths were recorded for the past 200 years or so. In the Labrador, Okak (see Table 1), Saglek and Labrador City experience frequent avalanche including fatalities. Avalanche also occur in several areas in Newfoundland such as St. Anthony, Port au Choix, Blue Mountain (see table), Lewis Hills, Tilt Cove, Fogo and others. Avalanche also affects roads and infrastructure.

Fisheries and Land Resources, Municipal Affairs and Environment, Natural Resources are departments looking into the problem as well as agencies such as Research and Development and Workplace Health and Safety.

Avalanche is also registered and can happen when the conditions allow for it in Nova Scotia namely in the Cape Breton area but also in places like Higgins Mountain (about 140 kilometers north of Halifax) and in the Wentworth Valley. Any slope that is more than 20 to 30 degrees of incline is subject to avalanche in that region or anywhere else in Eastern Canada including relatively small pieces of terrain like kid's tobogganing hills, which could produce an avalanche large enough to bury a child or even an adult if the terrain features are wrong according to The Canadian Avalanche Centre. The departments of Environment and Lands and Forestry and agencies like Emergency Management Office and Service Nova Scotia is trying to keep everybody safe.

CONCLUSION

Although Eastern Canada is not geographically as prone for fatal avalanche as Western Canada (claiming about 80 % of the death due to avalanche every year in Canada) and other parts of the world, it is still a growing problem with recent climate change and global warming. Provincial departments and agencies are constantly monitoring the conditions of snow, ice and weather in the winter time. Training, awareness, education, modern technology and more research will lead to fewer fatalities and injuries in the future while active in backcountry skiing or any other outdoor activity and keeping infrastructure such as highways, buildings, houses and others safer for everybody to use.

REFERENCES

Hétu B, Brown K, Germain D. 2008. L'inventaire des avalanches mortelles au Québec depuis 1825. *Proceedings of the 4th Canadian Conference on Geohazards*. PUL. Québec, 594 pp.

Isenor FM, Spooner IS, Wahl K, Liverman D, Smith J. 2005. Landslides and avalanches in Capr Breton Island, Nova Scotia, Canada. *Canadian Quaternary Association Conference*. Winnipeg, MB.

Jamieson B. 1997. Backcountry Avalanche Awareness. *Canadian Avalanche Association*. Revelstoke, BC. 42 pp.

Liverman DGE. 2018. Avalanches in Newfoundland and Labrador - an update. *Current Research (2018) Newfoundland and Labrador Department of Natural Resources Geological Survey*. Report 18-1: 59–69.

Toupin J. 2014. *Winter in Canada: Weird Facts, Records, Nostalgia, Hockey, Stats, Childhood Memories and More*. Folklore Publishing, Edmonton. 256 pp.

Toupin J. 2018. Avalanche in Western Canada: Who is in charge? *Proceedings of the 86th Conference of the Western Snow Conference.* Albuquerque, NM.

WEBSOURCES

Canadian Avalanche Center
Canadian Encyclopedia
Gouvernement du Québec
Avalanche Québec
Government of Newfoundland and Labrador
Government of Nova Scotia
Government of Ontario
Avalanche Canada

Synoptic Patterns Associated with Early and Late Onset of the Wet Season in Southern Peruvian Andes

TANIA ITA VARGAS[1], L. BAKER PERRY[1], HEATHER GUY[1], AND JOSEPH JONAITIS[1]

ABSTRACT

In the tropical Andes of southern Peru where a clear distinction between the wet and dry season exists, precipitation is an important atmospheric variable. Not only does it contribute to water supplies that are used in domestic, agriculture and industrial activities, but it is one of the most relevant factors in determining glacier mass balance. Glaciers located in the tropics are particularly sensitive to changes in precipitation. This is why the onset of the wet season is important because it interrupts the ablation period cause by low albedo and intense solar radiation at the end of dry season. Moreover, an early or late onset of the wet season could affect significantly the agriculture in southern Andes of Peru that depend heavily on rains during this period. In this study, daily precipitation observations from 1981 to 2017 are analyzed in order to define and identify the onset and end of the wet season as well as its duration and annual variability. Atmospheric fields from the ERA-Interim Reanalysis (0.75° and 6 hours spatial and temporal resolution) are analyzed to determine the configuration and position of the synoptic patterns associated with the onset and end of the wet season. Cases of early and late onset of the wet season are analyzed and compared to the climate period as well. Finally, the 1997-98 and 2015-16 strong El Niño years are analyzed to assess the influence of this phenomenon on an early or late onset of the wet season.

[1] Appalachian State University, Boone, NC, USA

Year-Round Estimation of Terrestrial Water Storage over Snow Covered Terrain via Multi-Sensor Assimilation of GRACE and AMSR-E

JING WANG[1], BARTON A. FORMAN[1], MANUELA GIROTTO[2], AND ROLF H. REICHLE[2]

ABSTRACT

The accuracy of terrestrial water storage (TWS) estimates is limited by a lack of observations and by inherent uncertainties in the model simulation. Although the Gravity and Recovery Climate Experiment (GRACE) has revolutionized large-scale remote sensing of the Earth's terrestrial hydrologic cycle, its coarse-scale (in space and time), vertically-integrated measure of TWS limits the applicability to smaller scale hydrologic applications. In order to enhance model-based estimates of TWS and its constituent components while effectively adding resolution (in space and time) to the coarse-scale TWS retrievals, a multi-variate, multi-sensor data assimilation framework is presented here that simultaneously assimilates gravimetric retrievals of TWS in conjunction with passive microwave (PMW) brightness temperature (Tb) observations over snow-covered terrain. The framework uses the NASA Catchment Land Surface Model (Catchment) and an ensemble Kalman filter (EnKF). A synthetic case study is presented for the Volga River Basin in Russia that compares model results with and without assimilation against synthetic observations of hydrologic states and fluxes. The AMSR-E/AMSR-2-only assimilation improved snow water equivalent (SWE) estimates. The GRACE-only assimilation improved TWS estimation but not always produced accurate estimates of SWE. The dual assimilation typically led to more accurate TWS and SWE estimates. The results demonstrate that GRACE TWS and AMSE-E can be jointly assimilated to produce improved TWS component estimate.

[1] University of Maryland, College Park, MD, USA
[2] NASA Goddard Space Flight Center, Greenbelt, MD, USA

Integration of a Spatiotemporal Subsampler for use in Observing System Simulation Experiments: Linking TAT-C with NASA LIS to Study Snow across Western Colorado

LIZHAO WANG[1], BARTON FORMAN[1], YONGHWAN KWON[1], AND SUJAY KUMAR[2]

ABSTRACT

When considering large-scale remote sensing of snow mass, it is desirable to achieve high spatial resolution with a short revisit period and a large signal noise ratio. However, it is difficult to achieve an optimal balance between these different variables as an improvement in one is often achieved at the expense of another. To explore the trade-off space between sensor design and orbital configuration, an Observing System Simulation Experiment (OSSE) is proposed for evaluating a suite of designs and configurations of a future satellite program.

In this study, we focused on the generation and evaluation of synthetic passive microwave brightness temperature observations over snow-covered terrain. To obtain the synthetic observations, our work consists of 3 steps: 1) generate synthetic observations of brightness temperature using a well-trained support vector machine to map "true" snow information from the NASA Land Surface System (LIS) into observational space, 2) use the Trade-space Analysis Tool for Constellations (TAT-C) to simulate the overpasses of a passive radiometer for a given orbital configuration, and 3) apply the space-time sub-sampler to mask out the non-observed portions of the study domain. In addition, we investigated how the properties of the sub-sampler would impact the error characteristics of the synthetic observations along with the trade-off between swath width, repeat frequency, spatial resolution, and signal-to-noise ratio. Results illustrate the quantitative interaction between the orbital configurations and desired observational characteristics, which can be useful for decision making based on the results from the OSSE to be pursued in a follow-on study.

[1] University of Maryland, College Park, MD, USA
[2] NASA Goddard Space Flight Center, Greenbelt, MD, USA

Towards the Development of a Hyper-Resolution High Mountain Asia-Land Data Assimilation System

YUAN XUE[1], YIWEN MEI[1], PAUL HOUSER[1], AND VIVIANA MAGGIONI[1]

ABSTRACT

The High Mountain Asia (HMA) region is one of the greatest mountain systems in the world, which contains one of the highest concentrations of snow (and glaciers) outside of the polar regions. However, an accurate representation of HMA snow states remains elusive due to the lack of dense and stable in-situ hydrometerology measurement networks as well as the spatio-temporal complexity of the land-atmosphere interactions triggered by the complex topography. In order to overcome the aforementioned limitations, this study developed a physically-based downscaling scheme for satellite-derived meteorological forcings using topography, roughness, and land cover observations (and/or reanalysis products). It is found that the downscaled products have higher values of Nash-Sutcliff index and correlation coefficient with the ground-based observations. A pattern-based comparison is also performed for the downscaled precipitation products with respect to the satellite precipitation products, suggesting fair correlation coefficient. In addition, this study explores the use of a hyper-resolution (1km) land data assimilation (DA) framework developed within the NASA Land Information System for quantifying snow water equivalent (SWE), snow depth, surface runoff, and top-layer soil moisture (0-10 cm) states. Three sets of DA experiments were conducted here, including satellite-derived snow cover map assimilation, landscape freeze/thaw state assimilation, and the joint assimilation of the two. The performance of the assimilation system is first evaluated across Indus River Basin from water year 2007 via comparisons to state-of-the-art SWE, snow depth, soil moisture products as well as available ground-based observations. In general, goodness-of-fit statistics were improved as a result of the assimilation. Future work will be extended to integrate the downscaled forcing product with the joint assimilation system in order to achieve a better understanding of the surface flux, snow/ice, and water balances in HMA.

[1] George Mason University, Fairfax, VA, USA

Estimating Snow Water Equivalent from a Combination of GPS and GRACE Observations over the Western United States

GAOHONG YIN[1], BARTON A. FORMAN[1], BRYANT D. LOOMIS[2], AND SCOTT B. LUTHCKE[2]

ABSTRACT

Snow plays a critical role in Western United States water supply – snow water equivalent (SWE) is one of the most important metrics in water resource forecasting. This study proposes the use of a combination of vertical displacement observations derived from a Global Positioning System (GPS) along with terrestrial water storage (TWS) retrievals derived from the Gravity Recovery and Climate Experiment (GRACE) mission in order to improve SWE estimates. After accounting for the effects of atmospheric loading, non-tidal ocean loading, and glacier isostatic adjustment (GIA) in the GPS observations, the remaining vertical displacement changes are driven predominately by terrestrial hydrologic processes, and thus, can be used to reflect the movement and redistribution of TWS and SWE.

For each GPS station, GRACE-based vertical displacement is acquired through the link between hydrological loadings and crustal deformation. A comparison of vertical displacement derived from GPS (after removing non-hydrological related loadings) with vertical displacement derived from GRACE in conjunction with SWE measurements from SNOW Telemetry (SNOTEL) is conducted for each GPS station in the study area. Correlation coefficients (R) between co-located GPS and SNOTEL, co-located GRACE and SNOTEL measurements, and co-located GPS and GRACE are provided. Over 75% stations provide a $R > 0.7$ between GPS-based and GRACE-based vertical displacement whereas a strong negative relationship can be detected between SWE and vertical displacement. The larger magnitude of variation in SWE corresponds to more variability in vertical displacement changes, which suggests potential for applying a combination of observations from both GPS and GRACE to better estimate regional SWE.

[1] University of Maryland, College Park, MD, USA
[2] NASA Goddard Space Flight Center, Greenbelt, MD, USA

Spatiotemporal Distribution of Snow in the High Mountain Asia and its Impact on Runoff

YEOSANG YOON[1,2], SUJAY V. KUMAR[1], DAVID M. MOCKO[1,2], ROBERT I. ROSENBERG[1,2], YONGHWAN KWON[3], BARTON FORMAN[3], AND BEN ZAITCHIK[4]

ABSTRACT

The melting of snow and glaciers in the High Mountain Asia (HMA) provides the water needs of approximately 1.3 billion people in the region. Increasing temperatures have large effects on the hydrologic cycle, influencing snowmelt, snowpack, streamflow, and water runoff, which can impact many aspects of water security. Most mountain regions, however, remain ungauged without in-situ measurement of precipitation or snowpack due to the complex terrain, and thus it is difficult to understand the regional water balance and assess how it might change in the future. In this study, we focus on characterizing the spatiotemporal patterns of snowpack states and fluxes over the last 30+ years (1980 – present) and assessing the relationship between snowmelt and runoff. Three different land surface models (i.e., Noah 3.3, Noah-MP 3.6, and Catchment F2.5) in the NASA Land Information System (LIS) is used to characterize the spatiotemporal pattern of snow. Combining information from satellite observations and the Modern-Era Retrospective analysis for Research and Applications, Version 2 (MERRA-2) is used to provide an effective way to develop spatially and temporally continuous estimates of changes. We evaluate our model estimates against satellite-derived data (e.g., MODIS snow cover fraction and GRACE total water storage anomaly) and reanalysis products (e.g., CMC and ERA-interim) using the NASA-developed Land Verification Toolkit. Using the model estimates, we examine spatiotemporal patterns of the snowmelt across HMA.

[1] NASA Goddard Space Flight Center, Greenbelt, MD, USA
[2] Science Applications International Corporation, McLean, VA, USA
[3] University of Maryland, College Park, MD, USA
[4] Johns Hopkins University, Baltimore, MD, USA

Fully Coherent Physical Model based on Analytical Method of Feynman Diagrams for Applications in Microwave Remote Sensing of Snow Cover

JIYUE ZHU[1], SHURUN TAN[1], LEUNG TSANG[1], AND SON V. NGHIEM[2]

ABSTRACT

Microwave remote sensing provides an important method to monitor the snow distributions globally. To accurately analyze microwave measurements, models of electromagnetic wave propagation, scattering, and emission of snow on ground or over sea ice are required. We used the bi-continuous model to characterize the microstructure of snow that contains information of snow grain size, grain clustering, and snow density, specific surface area (SSA), etc. The bi-continuous model can be used to generate microstructure of snow based on a computer algorithm. The microwave solutions for the bi-continuous medium (computer snow) are then calculated by the numerical method of solving Maxwell equations in 3-dimensions (NMM3D). The NMM3D is able to calculate the scattering properties (phase matrix, extinction coefficient, and effective permittivity) of snow samples, which are subsequently used in DMRT to obtain scattering coefficients and brightness temperatures. This approach is called partial coherent model (The RT is incoherent while NMM3D is fully coherent). Recently, we have also used NMM3D to solve the entire problem of snow cover fully coherently without using DMRT. In both DMRT methods and fully coherent methods, the NMM3D has high costs because of the Monte Carlo procedure of calculating the scattering solutions over many samples and the averages of the results calculated. Thus, we have revisited analytic method. With the single assumption of statistically homogenous medium, the Feynman diagram method is used to derive the Dyson's equations and the Bethe Salpeter equations from Maxwell equations. The method is exact and is fully coherent as Maxwell equations are the basis of the derivations. The Dyson's equations and the Bethe Salpter equations have the mass operator and the intensity operator, which contain infinite number of operators with the higher order operators representing higher order statistical moments of the heterogeneous permittivity. In the past, only the first order operator was used, giving the bi-local approximation with the mass operator approximated by the product of the pair function and the Green's function. The complexities of calculating higher correlation functions of medium prevent the use of higher order approximations beyond the bi-local approximation. We now return to the full Feynman diagrams because it is possible to calculate higher order correlation properties of the bi-continuous medium. The bi-continuous medium is generated by a Gaussian random process with the permittivity as a function of this Gaussian random process. Thus, higher-order statistical moments of the properties of the bi-continuous medium can be calculated. The Feynman diagram approach provides a fully coherent model without the costly Monte Carlo simulations. In this paper, we use the Feynman diagram method based on the layered medium Dyadic Green's function to solve the problem of layered bi-continuous medium. The mean field and scattering properties are calculated. The analytical Feynman diagram model is useful for applications to passive microwave radiometer polarimetry and active radar polarimetry and interferometry.

[1] University of Michigan, Ann Arbor, MI, USA
[2] Jet Propulsion Laboratory, California Institute of Technology, Pasadena, CA, USA

Session #4: NASA Snow Experiment (SnowEx) Campaign

Observing Snow in a Forested Environment: NASA's SnowEx Campaign Year 1

EDWARD KIM[1], CHARLES GATEBE[1,2], DOROTHY HALL[1,3], JERRY NEWLIN[4], AMY MISAKONIS[5], KELLY ELDER[6], HANS PETER MARSHALL[7], CHRIS HIEMSTRA[8], LUDOVIC BRUCKER[1,2], EUGENIA DE MARCO[1,4], DO HYUK KANG[1,3], CHRIS CRAWFORD[9], AND JARED ENTIN[10]

ABSTRACT

NASA's multi-year SnowEx airborne campaign is designed to collect measurements needed to enable algorithm development and to guide satellite mission trade studies. Snow community consensus is that a multi-sensor approach is needed to adequately address global SWE, combined with modeling and data assimilation to fill the gaps in space and time. Forests are a major confounding factor for the retrieval of SWE for as much as half of seasonally snow-covered terrestrial areas. Consequently, understanding how forest canopies influence remote sensing retrievals is important for planning any future snow satellite mission. SnowEx Year 1 (2016-17) focused on how best to combine and use various sensors to observe SWE and the snow energy balance in a forested environment. The Year 1 sites were Grand Mesa and the Senator Beck Basin, both in western Colorado, USA. This paper will describe the SnowEx Year 1 campaign, particularly the Feb 2017 winter deployment. Ground-based remote sensing and in situ data collection involved nearly 100 participants over three weeks. The airborne campaign included nine sensors on five aircraft. A broad suite of sensors, including active and passive microwave, and active and passive optical/infrared instruments, were deployed on aircraft, trucks, towers, snowmobiles, skis, and on foot to determine the sensitivity and accuracy of these potential satellite remote sensing techniques, along with models, to measure snow under a range of forest conditions. As data delivery is still in progress, this paper will provide an overview of the experiment design and execution, providing context for more focused papers. Selected examples of the data collected will be shown.

[1] NASA Goddard Space Flight Center, Greenbelt, MD, USA
[2] Universities Space Research Association, Columbia, MD, USA
[3] Earth System Science Interdisciplinary Center (ESSIC), University of Maryland, College Park, MD, USA
[4] ATA Aerospace, Greenbelt, MD, USA
[5] NASA Kennedy Space Center & Aerospace Corp., Greenbelt, MD, USA
[6] U.S. Forest Service, Fort Collins, CO, USA
[7] Boise State University, Boise, ID, USA
[8] U.S. Army Corps of Engineers Cold Regions Research and Engineering Laboratory, Fairbanks, AK, USA
[9] U.S. Geological Survey, Sioux Falls, SD, USA
[10] NASA Headquarters, Washington, DC, USA

Obtaining Reliable Retrieval of Snow Optical Properties from NASA's SnowEx Campaign Year 1

C.K. GATEBE[1,2], W. LI[3], N. CHEN[3], Y. FAN[3], R. POUDVAL[2,4], S. KHARBOUCHE[5], L. BRUCKER[1,2], AND K. STAMNES[3]

ABSTRACT

It is well known that the presence of snow on the ground affects the Earth's energy budget through its high albedo and thermal insulating properties and plays an important role in the global energy balance. Thus, knowledge of snow-covered area, snow water equivalent (SWE) and/or snow depth patterns, is needed in many practical applications involving snow (e.g. for water resource forecasting and simulations of snow related interactions with weather and climate). In 2017, five aircraft with a total of nine different sensors participated in the SnowEx campaign, carrying remote sensing sensors including active and passive microwave, and active and passive optical/infrared/thermal passive sensing techniques to determine the sensitivity and accuracy of potential satellite remote sensing techniques, along with models, to measure snow under a range of forest conditions. In this study, we will focus primarily on measurements by NASA's Cloud Absorption Radiometer (CAR) aboard the Naval Research Lab (NRL) Orion P-3C aircraft from February 16-22, 2017. The NRL P-3 flew the first science flight on February 16, based out of Colorado Springs, Colorado under clear sky. We will show results to demonstrate accurate and fast retrieval of the snow properties from CAR February 16 flight over Grand Mesa, Colorado. We will also show that CAR data can be a unique validation source for different snow models and satellite retrievals.

[1] Universities Space Research Association, Columbia, MD, USA
[2] NASA Goddard Space Flight Center, Greenbelt, MD, USA
[3] Stevens Institute of Technology, Hoboken, NJ, USA
[4] Science Systems and Applications, Lanham, MD, USA
[5] Mullard Space Science Laboratory, University College London, UK

Retrieval Algorithm of Snow Water Equivalent using SnowSAR and Scatterometer Backscatters with both Co- and Cross-Polarizations

JIYUE ZHU[1], SHURUN TAN[1], LEUNG TSANG[1], DO-HYUK KANG[2,3], AND EDWARD KIM[2]

ABSTRACT

A microwave volume scattering based retrieval algorithm is presented to invert a snow water equivalent (SWE) based on dual frequency (X- and Ku- bands—10 and 17 GHz) and single polarization (VV-polarization) radar measurements. The validation of this algorithm used 3 sets of airborne SnowSAR data (including 2011 and 2012 campaigns in Finland; 2013 campaign in Canada). The retrieval performance achieved a root-mean-square error (RSME) below 30 mm of SWE and a correlation coefficient above ~0.64. In the retrieval, the uncertainty associated with the lower boundary of the snowpack is reduced by predictions from scattering model of soil surface. Based on original dual band retrieval method, we also investigate the potential addition of lower Ku band (13 GHz) for SWE retrieval, which is less affected by this background scattering. Then, this combination with X-, Ku-, and lower Ku- bands is consistent with the SWE mission concept proposed by the Canadian Space Agency (CSA). While observations at cross-polarization provide sensitivity to handle thick snowpacks, this study shows the higher Ku band co-polarization backscatter may saturate when the snowpack is deeper than 4 meters. The proposed retrieval scheme thus forms a three-frequency and dual polarization radar SWE algorithm. Validations of the proposed algorithm is followed with the airborne SnowSAR data from Canada campaign 2013. The performance of the lower Ku band and cross-polarization for retrieval is cross-checked with the scatterometer data from Finland (ESA NoSREx; 2009-2013) and compared with previous results.

Keywords: bicontinuous dense media radiative transfer (Bic-DMRT), retrieval algorithm, snow water equivalent (SWE), lower Ku band (13 GHz), cross-polarization, SnowSAR, scatterometer

RESULTS

The SnowScat data were collected in Nordic snow radar experiment campaign (NoSREx) to support Earth Explorer mission CoReH2O. The data were measured by SnowScat at Sodankylä, Finland during the winter of 2010~2011. The instrument is three frequency, X-, low and high Ku-bands, dual polarization (VV & HV) ground-based scatterometer system.

In retrieval, a least-squares algorithm by a priori constrained by $\overline{\omega}_X$ is applied to fit the forward model results into radar observations. The cost function is given as:

[1] University of Michigan, Ann Arbor, MI, USA
[2] NASA Goddard Space Flight Center, Greenbelt, MD, USA
[3] Earth System Science Interdisciplinary Center (ESSIC), University of Maryland, College Park, MD, USA

$$F = \min_{\omega_X, \tau_X} \left\{ \frac{\left(\sigma_1^{obs} - \sigma_1^{model}(\omega_X, \tau_X)\right)^2}{2s_1^2} + \frac{\left(\sigma_2^{obs} - \sigma_2^{model}(\omega_X, \tau_X)\right)^2}{2s_2^2} + \frac{(\omega_X - \bar{\omega}_X)^2}{2s_{\omega X}^2} \right\} \quad (1)$$

where σ_1^{obs} and σ_2^{obs} are the two channel radar observations chosen for the class after subtraction background scattering, and σ_1^{model} and σ_2^{model} are the corresponding backscattering prediction by parameterized forward model. s_1 and s_2 are the expected error standard deviation of the measurements which is also regarded as the variance of speckle in a SAR image. In retrieval, we set $s_1 = s_2 = 0.5$ dB, $\bar{\omega}_X$ is the a priori value for albedo, and $s_{\omega X} = 0.1$ act as the standard deviation of the priori constraint. The contribution from each term in cost function is normalized by assuming a Gaussian distribution. ω_X and τ_X are the two parameters we need to solve from this equation. With cost function, we find the best fit ω_X and τ_X for each set of σ_1^{obs} and σ_2^{obs}.

In Figure 1 (a), (b) and (c), the performance of SWE retrieval for co-polarization Finland SnowScat data 2010-2011 based different frequency sets are inter-compared. In Table 1, we list the detail statistical results. We find retrievals based on all three frequency sets achieves RMSE less than 200mm and correlation above 0.85 which satisfy with the SCLP requirements. The combination of three frequency band retrieval gives best results showing that the low Ku band improves the retrieval especially for thin snow. In Figure 2 (a) and (b), the comparison of retrieval performance based on co-pol and cross-pol for thick snow is illustrated. In Table 2, we list the detail statistical results. The retrieval performance is improved for thick snow by introducing cross-polarization. RMSE and correlation are become better by using the cross-polarization.

Table 1. Performance of SWE retrieval for co-polarization Finland SnowScat data 2010-2011.

Co-pol channel	Correlation coefficient	RMSE (mm)	Bias (mm)
X and Ku high	0.854	14.63	6.81
Ku low and high	0.871	13.28	-5.41
Ku low and high for SWE < 60 mm X and Ku high for SWE > 60 mm	**0.913**	**11.54**	**1.66**

Table 2. Performance of SWE retrieval for co- and cross-pol Finland SnowScat data 2010-2011.

Channel	Correlation coefficient	RMSE (mm)	Bias (mm)
X and Ku co-pol	0.756	20.60	11.87
X and Ku cross-pol	**0.787**	**10.80**	**-2.66**

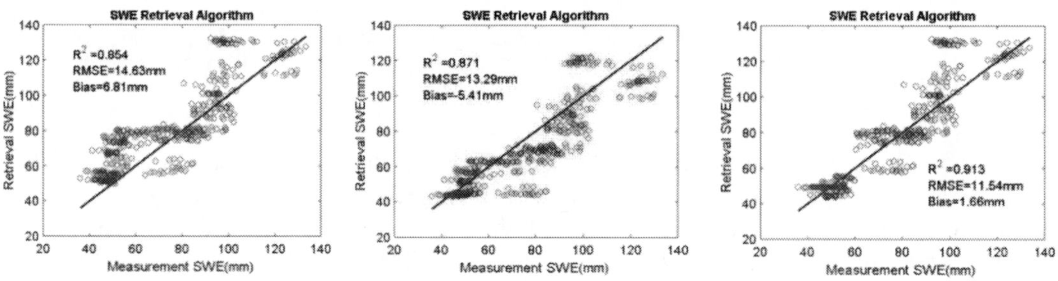

Figure 1. Comparison of SWE retrieved with SWE measured for co-polarization Finland SnowScat data 2010-2011. (a, left) retrieval based on co-pol X and high Ku band, (b, middle) retrieval based on co-pol two Ku band, and (c, right) retrieval based on combination of X-, low and high Ku-band.

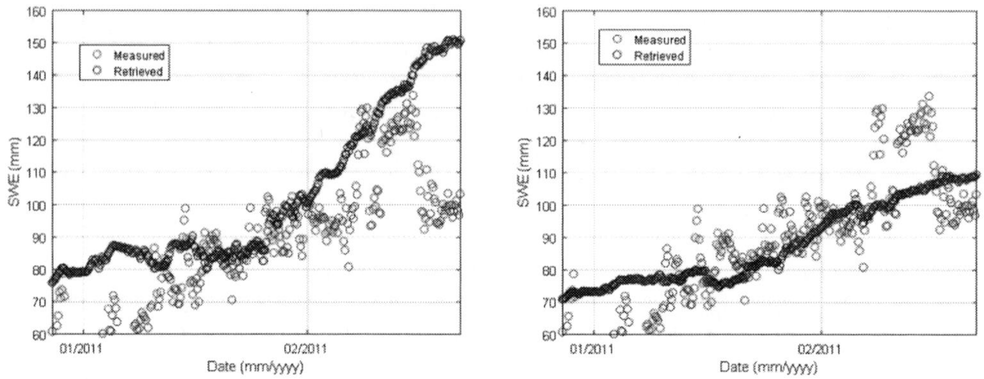

Figure 2. Comparison of SWE retrieved with SWE measured for Finland SnowScat data 2010 with SWE above 60 mm. (a, left) retrieval based on X and Ku-band co-pol, and (b, right) retrieval based on X and Ku band cross-pol.

REFERENCES

Zhu J, et al. 2018. Forward and Inverse Radar Modeling of Terrestrial Snow Using SnowSAR Data. *IEEE Transactions on Geoscience and Remote Sensing*, (in press).

Tan S, et al. 2015. Modeling both Active and Passive Microwave Remote Sensing of Snow Using Dense Media Radiative Transfer (DMRT) Theory with Multiple Scattering and Backscattering Enhancement. *IEEE Journal of Selected Topics in Applied Earth Observations and Remote Sensing*, **8**: 4418-4430.

Measurements of Snow Depth and Structure via Terrestrial LIDAR during SNOWEX

MANUEL A. SALGADO[1], ANDREW G. KLEIN[1], CHRISTOPHER A. HIEMSTRA[2], AND ARTHUR B. GELVIN[2]

ABSTRACT

The 2017 SNOWEX campaign, conducted on Grand Mesa, CO, combined aerial and terrestrial remote sensing measurements with extensive ground truthing in an effort to improve retrievals of snow water equivalent (SWE). Terrestrial LIDAR was deployed by researchers from the Cold Regions Research and Engineering Laboratory (CRREL) to scan multiple plots across the study area. These plots encompassed a variety of land cover types present including open grassland and forests of varying canopy density. Scans were initially conducted during snow free conditions and then again during the period of near peak snow depth. Snow depth analysis from the terrestrial LIDAR for four of the study areas shows spatial distributions correlated with the presence of vegetation and tree cover. Depth measurements in the forested areas were generally lower than the snow depths in adjacent open areas. Using the SNOWEX ground truth measurements to verify the LIDAR snow depth has proven problematic due to the low GPS positional accuracy (meters to tens of meters) for snow depth transects. However, the LIDAR datasets have centimeter level positional accuracy. Snow survey depth measurements in locations with high spatial variability of depth in snow depth have been found to vary substantially from the LIDAR measurements.

[1] Texas A&M University, College Station, TX, USA
[2] U.S. Army Corps of Engineers Cold Regions Research and Engineering Laboratory, Fairbanks, AK, USA

SnowEx 2017 Community Snow Depth Measurements: A Quality-Controlled, Georeferenced Product

L. BRUCKER[1,2], C. HEIMSTRA[3], H.-P. MARSHALL[4], AND K. ELDER[5]

ABSTRACT

Snow depth was one of the core ground measurements required to validate remotely-sensed data collected during SnowEx Year 1, which occurred in Colorado. The use of a single, common protocol was fundamental to produce a community reference dataset of high quality. Most of the nearly 100 Grand Mesa and Senator Beck Basin SnowEx ground crew participants contributed to this crucial dataset during 6-25 February 2017. Snow depths were measured along ~300 m transects, whose locations were determined according to a random-stratified approach using snowfall and tree-density gradients. Two-person teams used snowmobiles, skis, or snowshoes to travel to staked transect locations and to conduct measurements. Depths were measured with a 1-cm incremented probe every 3 meters along transects. In shallow areas of Grand Mesa, depth measurements were also collected with GPS snow-depth probes (a.k.a. MagnaProbes) at ~1-m intervals. During summer 2017, all reference stake positions were surveyed with <10 cm accuracy to improve overall snow depth location accuracy.

During the campaign, 193 transects were measured over three weeks at Grand Mesa and 40 were collected over two weeks in Senator Beck Basin, representing more than 27,000 depth values. Each day of the campaign depth measurements were written in waterproof field books and photographed by National Snow and Ice Data Center (NSIDC) participants. The data were later transcribed and prepared for extensive quality assessment and control. Common issues such as protocol errors (e.g., survey in reverse direction), notebook image issues (e.g., halo in the center of digitized picture), and data-entry errors (sloppy writing and transcription errors) were identified and fixed on a point-by-point basis. In addition, we strove to produce a georeferenced product of fine quality, so we calculated and interpolated coordinates for every depth measurement based on surveyed stakes and the number of measurements made per transect.

The product has been submitted to NSIDC in csv format. To educate data users, we present the study design and processing steps that have improved the quality and usability of this product. Also, we will address measurement and design uncertainties, which are different in open vs. forest areas.

[1] NASA Goddard Space Flight Center, Greenbelt, MD, USA
[2] Universities Space Research Association, Columbia, MD, USA
[3] U.S. Army Corps of Engineers Cold Regions Research and Engineering Laboratory, Fairbanks, AK, USA
[4] Boise State University, Boise, ID, USA
[5] U.S. Forest Service, Rocky Mountain Research Station, Fort Collins, CO, USA

Session #5: Snow Water Equivalent and Watershed Hydrology

Mapping Snow Mass in the European Alps Using Sentinel-1 Radar Observations

H. LIEVENS[1,2,3], R.H. REICHLE[2], M. GIROTTO[2,4], L. BRUCKER[2,4], E. KIM[2], C. MARTY[5], T. JONAS[5], M. OLEFS[6], M. DUMONT[7], D. VERFAILLIE[7], J. SCHOEBER[8], AND G.J.M. DE LANNOY[3]

ABSTRACT

Seasonal snow packs provide the water supply for more than one-sixth of the world's population and have major impacts on climate. However, the spatio-temporal variability of snow is still poorly understood, particularly in mountainous areas, owing to a lack of high-resolution satellite observations and robust snow retrieval algorithms. We demonstrate the ability of the Sentinel-1 mission to monitor snow water equivalent (SWE) at 1-km resolution over the European Alps for 2015-2017. The SWE retrievals rely on a change detection algorithm using the cross-polarization ratio, or the ratio of cross-polarization to co-polarization backscatter. We show that this ratio is sensitive to SWE throughout winter. For dry snow conditions, Sentinel-1 SWE retrievals and in situ measurements from 588 sites across the Alps correlate to 0.8 over time and 0.68 over space, with a root-mean-square difference (RMSD) of 0.06 m. The SWE retrievals outperform globally available 9-km model simulations, which have a spatial correlation of only 0.28 vs. *in situ* measurements. The Sentinel-1 SWE retrievals also match regional, 1-km resolution snow model simulations over Austria and Switzerland, with correlations of 0.81 in time and up to 0.91 in space. Correlations are slightly lower over densely forested areas and somewhat lower over high-elevation areas with high simulated SWE during wet snow conditions. The results highlight the potential of Sentinel-1 observations to provide regional snow estimates at an unprecedented (1-km) resolution in mountainous areas.

[1] Ghent University, Ghent, Belgium
[2] NASA Goddard Space Flight Center, Greenbelt, MD, USA
[3] KU Leuven, Leuven, Belgium
[4] Universities Space Research Association, Columbia, MD, USA
[5] WSL-Institute for Snow and Avalanche Research SLF, Davos, Switzerland
[6] ZAMG-Zentralanstalt für Meteorologie und Geodynamic, Vienna, Austria
[7] CNRM/CEN, Météo-France – CNRS, Grenoble, France
[8] TIWAG-Tiroler Wasserkraft AG, Innsbruck, Austria

Investigating the 2009 Red River of the North Snowmelt Flood using Enhanced Resolution Passive Microwave Data

MARISSA TORRES[1], MARINA REILLY-COLLETTE, AND CARRIE VUYOVICH

EXTENDED ABSTRACT

Keywords: passive microwave, snow water equivalent, brightness temperature, melt signal, Red River of the North

A newly processed NASA MEaSUREs dataset, the Calibrated Enhanced-Resolution Passive Microwave Brightness Temperature (CETB) provides higher-resolution passive microwave data at frequencies used to observe snow. These data were developed for SMMR, SSM/I-SSMIs and AMSR-E passive microwave observations, potentially providing a high-resolution record of snow mass and snowmelt indicators in the Red River of the North (RRN) basin. This work presents a case study for the RRN focusing on the March 2009 snowmelt-driven flood. The CETB data were evaluated along with hydro-meteorological data for the event to assess potential for identifying snowpack ripeness and melt onset. SWE was estimated from the CETB data using a simple empirically-based algorithm, and compared to coarser-resolution passive microwave SWE products and SNODAS modeled SWE estimates.

The Red River of the North basin is a hydrologically complex area situated in the northern Midwest of the United States. The Red River has an extensive history of flooding, primarily due to its geographical features. Positioned in a low-lying valley, the Red River mainstem flows northward over a very shallow gradient (2.64 m per km (1/2 ft per mi); Schwert, 2003). Increases in river water surface elevation can cause widespread flooding along the entire basin. High river discharge resulting from spring thaw poses the greatest threat to communities.

The flooding event of March 2009 was the second worst in the RRN modern history, reaching a peak stage of 12.5 m (41 ft) in Fargo, ND. The severity of the flood was due to a combination of high soil moisture (near 100%) from the previous fall, above normal winter snowpack, and increased rainfall with unseasonably warm temperatures in late March (Macek-Rowland and Gross, 2011). Flood levels persisted for two months after peak stage, causing up to $55M USD in damage to the region (National Weather Service, 2010).

The National Weather Service (NWS) North Central River Forecast Center (NCRFC) uses the Community Hydrologic Prediction System (CHPS) to forecast snowmelt and streamflow in the Red River of the North, based on modeled and observed estimates of snow water equivalent (SWE). While the enhanced resolution passive microwave data are not currently available real-time to support flood forecasting, we investigate the potential capability for these data to identify melt at a higher resolution than previously available. In this study, we compute SWE from the CETB passive microwave product using a simple algorithm. Comparing this product with the coarser resolution

[1] U.S. Army Corps of Engineers Cold Regions Research and Engineering Laboratory, Hanover, NH, USA

microwave SWE, modeled SWE, and in situ observations (Table 1), we evaluate the evolution of the March 2009 snow ripening and melt event, and determine if a clear melt signal can be detected.

The CETB is a NASA MEaSURES product on the EASE 2.0 grid that produces a long-term record of higher resolution (3.125 km) passive microwave brightness temperature data. Using a combination of the 36 and 18 GHz horizontal bands from AMSR-E, we can compute SWE with an empirical algorithm originally developed for SMMR (Chang, 1987) and later modified for SSM/I (Armstrong & Brodzik, 2001). Melt signals in the passive microwave data were evaluated as a sharp decrease (e.g. loss) of SWE signal over a basin-averaged time series (Schroeder et al, 2018). It is well known that a loss of SWE signal occurs during periods of wet snow; however, we believe these "bad returns" could be an indicator for melt signals.

Table 1. Sources of spatial data (modeled and remotely sensed) and ground observations used in this study

		Spatial (Modeled/Remotely Sensed)				
Product	Data Type	Spatial Coverage	Spatial Resolution	Temporal Coverage	Temporal Resolution	Source
SNODAS	SWE	CONUS	1 km	2003-2018	Daily	NOAA
AMSR-E	SWE	Global	25 km	2002-2011	Daily	NASA
SSM/I	SWE	Global	25 km	1987-2018	Daily	DMSP
CETB	SWE	Global	3.125 km	1978-2017	12 hr	NASA
PRISM	Temp, Precip	CONUS	4 km	1981-2017	Daily	Oregon State
MODIS	SCA (w/ temporal filter)	Global	1 km	2000-2018	Daily	NASA
		In situ Observations				
CDO	Precip, Snow fall	CONUS	Point	Varies	Daily	NOAA
ISD	Precip	CONUS	Point	Varies	Hourly	NOAA
NDAWN	Temp, Precip	North Dakota	Point	1989-2018	Hourly	NDSU
Stream Flow	Discharge	CONUS	Point	Varies	Daily	USGS

Results from the March 2009 analysis show clear signal changes in the passive microwave data in the days preceding the flood, characteristic of the effects of snowpack ripening on the microwave signal (Figure 1). This signal change is particularly evident along the valley of the RRN mainstem and some of the south-eastern subbasins. Further, there is a distinct spatial relationship between air temperature and loss of passive microwave SWE estimates. Discharge data in several unregulated RRN subbasins were used to independently evaluate the timing of snow-melt as indicated by the sharp decrease in the CETB data (Figure 2). River discharge is shown to increase significantly prior to rainfall precipitation over the basin, most likely as a result of snowmelt runoff.

An initial observation is that despite the extreme flatness of the Red River of the North basin and lack of significant elevation change, the CETB data indicates heterogeneous behavior in the spatial distribution of the initial melt onset which may provide valuable hydrologic information.

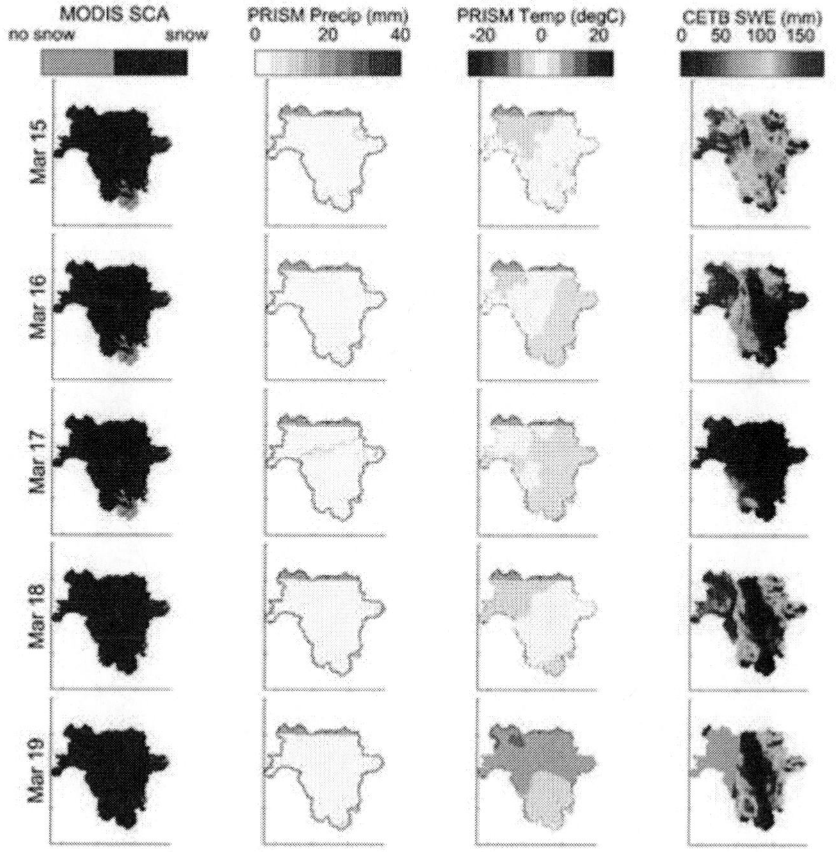

Figure 1. Spatial maps of SCA (MODIS) , precipitation and temperature (PRISM), and SWE (CETB) in the RRN from March 15 to March 19, 2009.

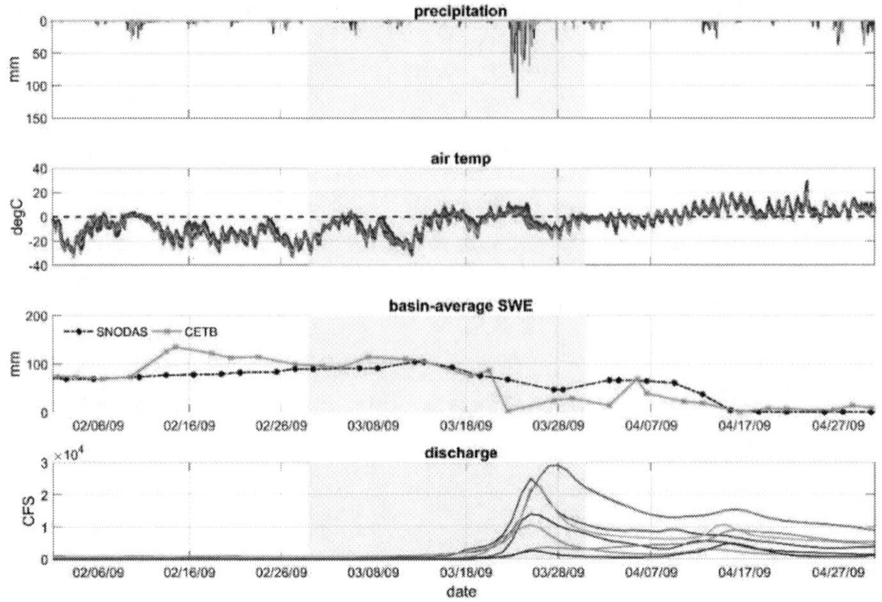

Figure 2. Time series of precipitation (ISD), air temperature (NDAWN), SWE (SNODAS & CETB), and discharge (USGS) in the RRN from February to April 2009; the month of March is highlighted in blue.

REFERENCES

Armstrong RL, Brodzik MJ. 2001. Recent Northern Hemisphere Snow Extent: A Comparison of Data Derived from Visible and Microwave Satellite Sensors. *Geophysical Research Letters*, **28**(19): 3673-3676.

Chang ATC, Foster JL, Hall DK. 1987. Nimbus-7 Derived Global Snow Cover Parameters. *Annals of Glaciology* **9**: 39-44.

Macek-Rowland KM, Gross TA. 2011. *2009 Spring floods in North Dakota, western Minnesota, and northeastern South Dakota*. U.S. Geological Survey Scientific Investigations Report 2010–5225, 41 pp.

National Weather Service (NWS). 2010. *Flood Damages Suffered in the United States During Water Year 2009*. Annual Flood Loss Summary Reports. National Oceanic and Atmospheric Administration (NOAA).

Schroeder R, Kraatz S, Jacobs JM, Vuyovich CM, Olheiser C, Connelly B, DeWeese MM. 2018. Detection of snowmelt signals for improving snowmelt flood forecast in the Red River basin of the North. 75th Eastern Snow Conference, June 5-8, College Park, MD.

Schwert DP. 2003. *A geologist's perspective on the Red River of the North: history, geography, and planning/management issues*. Proceedings 1st International Water Conferences, Red River Basin Institute, Moorhead, MN.

Sub-Pixel Variability of the Measured Ice or Snow Pack Thicknesses using Wideband Autocorrelation Radiometer

MOHAMMAD MOUSAVI[1], ROGER DE ROO[2], KAMAL SARABANDI[1], AND ANTHONY W. ENGLAND[3]

ABSTRACT

The seasonal terrestrial snowpack is an important source of water for many parts of the globe. Snow's high albedo, relative to the terrain in the absence of snow, is an important driver of Earth's energy balance, and long term changes to the statistics of the snowpack's properties are both a consequence and a cause of climate change. In most remote sensing applications, the gross parameter of the target, such as snow depth and snow water equivalent (SWE), are often the parameters of interest. A novel and new passive microwave remote sensing technique, known as wideband autocorrelation radiometry (WiBAR), offers a direct method to remotely measure the microwave propagation time difference of multipath microwave emission from low-loss layered surfaces such as a dry snowpack and freshwater Lake Icepack. The microwave propagation time difference through the pack yields a measure of its vertical extent.

The presence of variable pack thicknesses in a footprint of the radiometer's antenna will add more complexity to the retrieved time delay. This issue is more severe for WiBAR on airborne and space-borne platforms since the footprint will be large and contains more variable thicknesses. We present a simple forward model to include the variable thicknesses in one pixel and derive the system requirements needed to observe sub-pixel variability in the measured pack thickness. An X-band instrument fabricated from components-off-the-shelf (COTS) measured the thickness of freshwater lake ice at the University of Michigan Biological Station. Sub-pixel variability of 3 cm is demonstrated at incident angle of about 70°.

[1] Electrical Engineering and Computer Science Department, University of Michigan, Ann Arbor, MI, USA
[2] Climate and Space Sciences and Engineering Department, University of Michigan, Ann Arbor, MI, USA
[3] College of Engineering and Computer Science, University of Michigan, Dearborn, MI, USA

Wet Snow Detection from Radarsat-2 Images in Nunavut, Canada

YULIA ANTROPOVA[1], ALEXANDER S. KOMAROV[2], MURRAY RICHARDSON[1],
KOREEN MILLARD[1], AND KEEGAN SMITH[1]

ABSTRACT

The 2017 SNOWEX campaign, conducted on Grand Mesa, CO, combined aerial and terrestrial remote sensing measurements with extensive ground truthing in an effort to improve retrievals of snow water equivalent (SWE). Terrestrial LIDAR was deployed by researchers from the Cold Regions Research and Engineering Laboratory (CRREL) to scan multiple plots across the study area. These plots encompassed a variety of land cover types present including open grassland and forests of varying canopy density. Scans were initially conducted during snow free conditions and then again during the period of near peak snow depth. Snow depth analysis from the terrestrial LIDAR for four of the study areas shows spatial distributions correlated with the presence of vegetation and tree cover. Depth measurements in the forested areas were generally lower than the snow depths in adjacent open areas. Using the SNOWEX ground truth measurements to verify the LIDAR snow depth has proven problematic due to the low GPS positional accuracy (meters to tens of meters) for snow depth transects. However, the LIDAR datasets have centimeter level positional accuracy. Snow survey depth measurements in locations with high spatial variability of depth in snow depth have been found to vary substantially from the LIDAR measurements.

[1] Carleton University, Ottawa, ON, Canada
[2] Environment and Climate Change Canada, Ottawa, ON, Canada

Reconsidering the Utility of the April 1st Snow Water Equivalent Metric

KEITH N. MUSSELMAN[1], JOHN BERGGREN[2], JULIE VANO[3], NANS ADDOR[4], AND NOAH P. MOLOTCH[1]

ABSTRACT

Measurements of April 1st snow water equivalent (SWE) are regularly used as a metric of the total meltwater available for spring runoff. This date generally coincides with the transition from winter accumulation to spring melt. With increasing snowpack information from models, the magnitude and date of maximum SWE as a spatial aggregate (i.e., basin total) can now be better characterized. Yet, the operational use of spatial metrics (i.e., SWE maps) by water managers remains limited despite an uptick in spatial SWE products and climate change impacts such as declines in snowpack volumes and trends toward earlier melt. We examine how basin-wide snowpack volume and its seasonal maximum varies historically across western North America and how those metrics may respond to end-of-century warming. Here we compared the date of maximum annual SWE measured at 979 snow pillows in western North America to the date simulated by the Weather Research and Forecasting (WRF) model run over 2000-2013 at 4 km resolution. When aggregated to river basins, the date of maximum SWE differed greatly from station-based dates, and only coincided with April 1st in a fraction of the basins. Further, in the warmer scenario, the WRF-simulated date of maximum SWE occurs on average one month earlier, which further compromised the adequacy of April 1st SWE to characterize the potential spring runoff. In this presentation, we discuss why water managers value April 1st SWE and evaluate alternative metrics that could provide similar information under present and future climate.

[1] INSTAAR, University of Colorado, Boulder, CO, USA
[2] Center for Science and Technology Policy Research, Boulder, CO, USA
[3] National Center for Atmospheric Research, Boulder, CO, USA
[4] University of East Anglia, Norwich, UK

Session #6: Snow and Ice Remote Sensing

Assessment of Uncertainties in the New MODIS Cloud-Gap Filled Snow Maps

DOROTHY K. HALL[1,2], GEORGE A. RIGGS[2,3], AND NICOLO E. DIGIROLAMO[2,3]

EXTENDED ABSTRACT

Keywords: snow mapping, MODIS, Terra, Aqua, VIIRS, cloud gap filling

MODerate resolution Imaging Spectroradiometer (MODIS) cryosphere products that have been available since the launch of the Terra MODIS in 2000 and the Aqua MODIS in 2002 include snow-cover extent and daily snow albedo, as well as extent and ice-surface temperature (IST) of sea ice, and IST of the Greenland Ice Sheet. Reprocessing, from Collection 5 (C5) to Collection 6 (C6) and Collection 6.1 (C6.1), has led to improvements in the MODIS snow-cover extent, daily snow albedo and IST standard data products (Riggs et al., 2016; Riggs et al., 2017). Work is ongoing to evaluate and document uncertainties in the C6 suite of cryosphere products. Here, we focus on a new product, MOD/MYD10A1F, which is a daily, cloud-gap filled (CGF) snow-cover map (Hall et al., 2010).

Researchers have used various methods to reduce cloud cover in the MODIS snow maps. One common method is to combine Terra (MOD10A1) and Aqua (MYD10A1) snow maps for a given day. This takes advantage of the fact that the Terra and Aqua satellite overpasses occur at different times of the day and, since clouds move, oftentimes more snow cover or non-snow-covered land cover can be imaged and mapped by the snow-cover mapping algorithm, as compared to using the Terra or Aqua snow map alone. However, this method of cloud clearing is of limited utility because typically it does not allow a great deal more snow or non-snow-covered land to be observed/mapped on a given day.

As part of the C5 and C6 snow product suite, 8-day maximum snow-cover maps (MOD/MYD10A2) have been available to provide greatly-reduced cloud cover for the MODIS snow maps, but are available only once every eight days, and often retain some cloud cover which is not ideal for many users. Nevertheless, these snow maps have been useful in numerous research studies.

Cloud-gap filling is another common method that provides cloud-reduced or cloud-free snow maps. For C6 we have adopted this technique to enable fewer clouds, and more snow (or snow-free land), to be mapped on a given day (Figure 1). (It is expected that the new CGF maps will replace the MOD/MYD10A2 (8-day maps)). Though cloud-gap filling provides a cloud-free snow map every day, the accuracy of any given decision at the pixel level depends in part on the age of the snow/no-snow decision.

---

[1] Earth System Science Interdisciplinary Center (ESSIC), University of Maryland, College Park, MD, USA
[2] NASA Goddard Space Flight Center, Greenbelt, MD, USA
[3] Science Systems and Applications (SSAI), Lanham, MD, USA

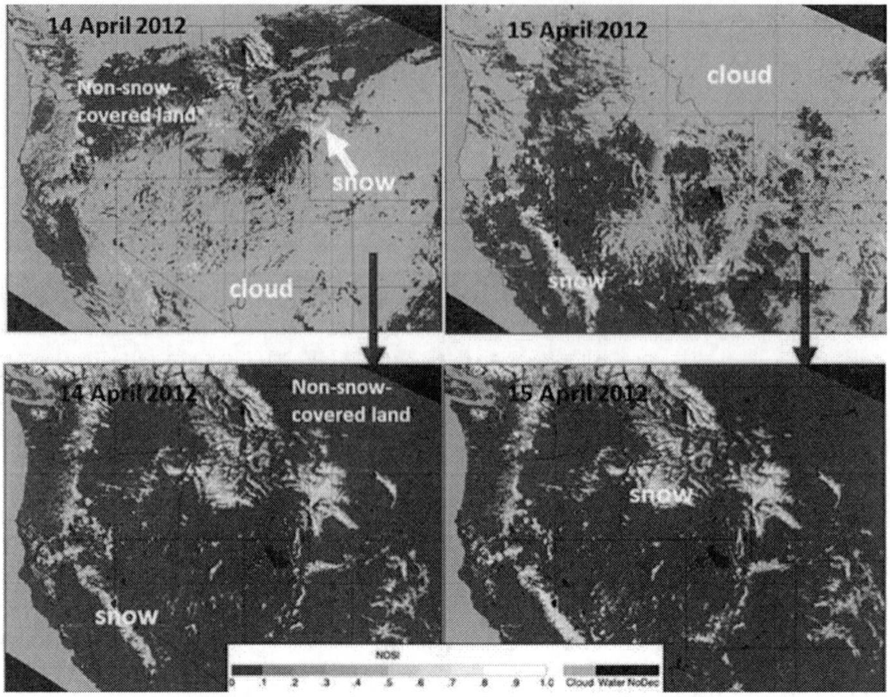

Figure 1. Examples of the MOD10A1 and the new MOD10A1F MODIS snow maps. Top row: Note the extensive cloud cover on 14 and 15 April 2012 on the MOD10A1 snow maps. Bottom row: The corresponding MOD10A1F, or cloud-gap filled (CGF) maps. Non-snow-covered land is shown in green.

A new feature of the C6 product suite provides the snow maps as normalized-difference snow index (NDSI) maps instead of fractional-snow cover (FSC) maps as was done in C5. This has the advantage of allowing a user to more-accurately determine FSC in their particular study area by applying a simple algorithm to derive FSC from the NDSI. The C5 global FSC algorithm was useful for estimating FSC for very large areas, but of more-limited utility for specific study areas.

For a time series of CGF snow maps extending from 1 February through 30 April 2012, comparisons have been conducted with other snow-cover maps such as those developed from the Visible Infrared Imaging Radiometer Suite (VIIRS) and from the National Ice Center's Interactive Multispectral Snow and Ice Mapping System (IMS), and with true-color images from satellites such as the Terra MODIS, Aqua MODIS, VIIRS and Landsat. Validation of snow-cover products with in situ meteorological-station data has also been conducted for some study areas.

Quality-assurance (QA) information can be exploited to help a user understand product uncertainties and has been available in all previous versions of the MODIS snow-cover mapping algorithms. In C6, a new QA map showing cloud-persistence count (CPC) on a per-pixel basis is available to go along with each CGF snow map. For example, for 19 March 2012, when CPC=0 this means that the reported NDSI value for that pixel was acquired on 19 March 2012. When CPC=1 this means that the reported NDSI value is one day old, hence it was acquired on 18 March, and so on (Figure 2). A user can decide how far back in time they would like to use a snow observation and can easily develop an appropriate CGF map for their application.

Another useful method in C6 for analysis of uncertainties is the ability to remove the cloud mask, MOD/MYD35, that is an input to the MODIS snow-mapping algorithm. In C6, the cloud mask can be 'removed' so that a user can visually see what is beneath the clouds. Sometimes the shape of features beneath clouds can be indicative of snow cover on terrain, and a user might make the assumption or decide that the cloud mask algorithm failed to properly distinguish snow and clouds and thus incorrectly mapped cloud when the feature was really snow. The ability to remove the cloud mask may allow a user to view more snow cover but must be used with a great deal of caution.

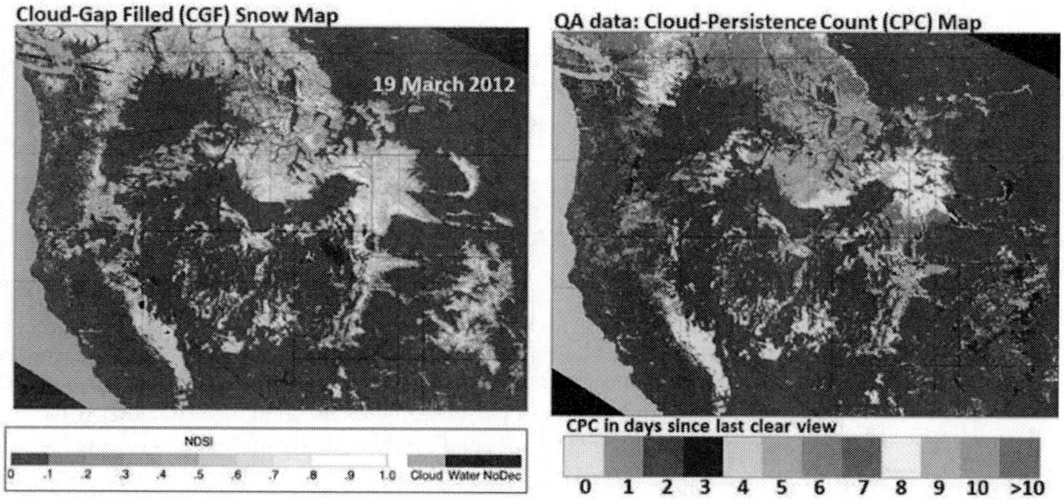

Figure 2. Left - Cloud-gap filled (CGF) MOD10A1F map for 19 March 2012. Right – Cloud-Persistence Count (CPC) map from the Quality Assurance (QA) information. Each CGF map comes with a CPC map so that a user may determine the age of the snow observation on a per-pixel basis.

Preliminary results demonstrate the great utility of the MODIS CGF maps as long as uncertainty information is understood by a user. The CGF product is scheduled to be available in the summer of 2018.

REFERENCES

Hall DK, Riggs GA, Foster JL and Kumar SV. 2010. Development and evaluation of a cloud-gap-filled MODIS daily snow-cover product. *Remote Sensing of Environment*, **114**: 496-503. doi:10.1016/j.rse.2009.10.007.

Riggs GA, Hall DK, Román MO. 2016. MODIS snow products user guide for Collection 6 (C6). Available at: *https://modis-snow-ice.gsfc.nasa.gov/?c=userguides*. Last access: 8 March 2017.

Riggs GA, Hall DK, Román MO. 2017. Overview of NASA's MODIS and Visible Infrared Imaging Radiometer Suite (VIIRS) snow-cover Earth system data records. *Earth System Science Data*, **9**: 765-777.

Global Snow Zone Maps and Trends in Snow Persistence 2001-2016

JOHN C. HAMMOND[1], FREDDY A. SAAVEDRA[2,3], AND STEPHANIE K. KAMPF[4]

ABSTRACT

Seasonal snow is a critical component of the surface energy balance and hydrologic cycle, yet global maps of seasonal snow boundaries are not readily available. Snow persistence (SP), the fraction of a year that snow is present on the ground, is an easily globally observed snow metric that can be used to map snow zones globally. Here we map snow zones across the globe using SP calculated from the MODIS10A2 product; evaluate how SP relates to precipitation, temperature, and climate indices, and examine trends in annual SP for 2001-2016. In the northern hemisphere, intermittent, seasonal, and permanent snow zones occupy a far greater percent (63%) of the land surface than in the southern hemisphere (< 5%) where the low snow zone dominates (> 95%). SP is most variable from year to year near the snow line, which has a relatively consistent decrease in elevation with increasing latitude across all continents. Areas with decreasing SP trends cover 5.8% of snow zone areas, whereas those with increasing trends cover 1.0% of this area. The largest areas of declining SP are in the seasonal snow zones of the northern hemisphere. Trend patterns vary within individual regions, with elevation, and on windward-leeward sides of the mountains. This study supplies a framework for comparing snow between regions, highlights areas with snow changes, and can facilitate analyses of why snow changes vary within and between regions.

[1] Department of Geosciences, Colorado State University, Fort Collins, CO, USA
[2] Departmento de Ciencias Geográficas, Universidad de Playa Ancha, Valparaíso, Chile
[3] Centro de Estudios Avanzados, Universidad de Playa Ancha, Viña del Mar, Chile
[4] Department of Ecosystem Science and Sustainability, Colorado State University, Fort Collins, CO, USA

Monitoring Ice Phenology of Small Ponds and Lakes using Sentinel-1 and Cloud-Based Detection Algorithms

GRANT E. GUNN[1], ERIN BUNTING[1], AND DI YANG[2]

ABSTRACT

Lake ice phenology (freeze-up, break-up dates, total days ice-covered) has been tracked using active microwave scatterometers and SAR sensors with varying degrees of success, typically limited by spatial resolution and temporal resolution, respectively. With the launch of Sentinel-1 A/B, consistent SAR acquisitions data are freely available, increasing the accessibility and operational utility of the constellation of high-resolution C-band SAR observations to observe the evolution of ice cover over freshwater lakes and ponds in northern environments.

This study introduces the development and preliminary results of an automated approach to track ice phenology using C-band acquisitions for freshwater lakes on the North Slope of Alaska. The algorithm utilizes the catalogue of Sentinel-1 A/B acquisitions available to Google Earth Engine to produce an estimation of the dates thermodynamic regime shifts on a per-pixel basis, producing intra-lake ice phenology at a minimum spatial resolution of 40 x 40 meters. The development of automated cloud-based algorithms to track physical characteristics of landcover is a burgeoning area of research development. With the movement towards an increase in temporal observations with SmallSATs, as well as the upcoming launch of the RADARSAT Constellation Mission slated to increase the data volume available to researchers, so do the computational requirements for algorithm development. This presentation presents an initial step to utilize cloud-based processing for the identification of physical parameters of the cryosphere.

[1] Michigan State University, East Lansing, MI, USA
[2] University of Florida, Gainesville, FL, USA

Using a Convolutional Neural Network to Classify Ice/Water Conditions from Different C-Band SAR Platforms in the Arctic

BENOIT MONTPETIT[1], BENJAMIN DESCHAMPS[2], STEPHEN HOWELL[3], DAVID A. CLAUSI[4], MOHAMMAD JAVAD SHAFIEE[4], JASON DUFFE[1], AND DEAN FLETT[2]

ABSTRACT

Arctic air temperatures are increasing by approximately twice the amount as the global mean. The response of the sea ice to warmer temperatures has major impacts on the climate, ecosystems, communities, wildlife and economy in the Arctic. The ice-air ocean-air interactions have major impacts on atmospheric processes and snowpack properties in the Arctic. One of the main applications of the Canadian RADARSAT Program is to monitor the sea ice conditions in Canadian waters. The C-Band synthetic aperture radar (SAR) sensor of RADARSAT-1 was designed for optimal monitoring of sea ice conditions in the Canadian Arctic.

The Canadian Ice Service (CIS) of Environment and Climate Change Canada has been providing sea ice charts since the 1970s. Modern charts provided by the CIS have been manually digitized mostly from RADARSAT-1/2 imagery, since 1997. With the increasing amount of C-Band SAR sensors like Sentinel-1A and -1B of the European Space Agency and the soon to be launched RADARSAT Constellation Mission (RCM), it is becoming more difficult to fully exploit of all this available data with traditional charting methods.

Machine learning algorithms have proven useful in automating sea ice feature extraction but have proven to be very dependent on many parameters like time of year, viewing geometry, sensor and image characteristics and ice regime. These methods are promising but prove to be very computationally costly and complex to implement. Deep learning methods on the other hand have the potential to use all this data and learn, through proper training, how to classify sea ice conditions no matter the sensor, parameterization, or location.

The convolutional neural network (CNN) proposed in this study is potentially the closest deep learning method to visual digitization since it is capable of identifying features from imagery at different scales similar to what ice analysts currently do visually to produce the ice charts. The proposed CNN has been trained and validated on 10 years of dual-pol RADARSAT-2 ScanSAR images. It was also tested on Sentinel-1A and -1B EW mode as well as simulated RADARSAT Constellation Mission images with similar accuracies as RADARSAT-2. This method is proving very promising for exploiting and integrating large amounts of imagery from C-band and other SAR sensors, irrespective of the acquisition parameters and target characteristics.

---

[1] National Wildlife Research Centre, Environment and Climate Change Canada, Ottawa, ON, Canada
[2] Canadian Ice Services, Environmental and Climate Change Canada, Ottawa, ON, Canada
[3] Climate Research Division, Environment and Climate Change Canada, Toronto, ON, Canada
[4] University of Waterloo, Waterloo, ON, Canada

Uncertainty in Future Changes in Snowpack and Rain-on-Snow Events in the U.S. Northern Great Plains using High-Resolution Climate Models

RACHEL R. MCCRARY[1], MELISSA BUKOVSKY[1], AND LINDA O. MEARNS[1]

ABSTRACT

In the Northern Great Plains region of the United States, spring and summer flooding is often correlated with high antecedent winter snowpack conditions. For example, in the Missouri River Basin and the Red River of the North, most of the major historic flood events have corresponded with rapid snow melt or rain-on-snow events. Snow accumulation and melt dynamics are almost certainly going to change as regional temperatures warm over time, shifting the magnitude and seasonality of flood risk in this region. This study explores how anthropogenic climate change is projected to influence snowpack characteristics (timing, amount, and distribution) and the frequency and intensity of rain-on-snow events in the Northern Great Plains using a suite of dynamically downscaled, high-resolution (12km, 25km, and 50km) climate change simulations from the North America Coordinated Regional Climate Downscaling Experiment (NA-CORDEX) and the Framework for Assessing Climate's Energy-water-land nexus using Targeted Simulations (FACETS) ensembles. Analysis includes an evaluation of the historical climate simulations as well as estimation of future changes for snowpack dynamics and rain-on-snow events.

[1] National Center for Atmospheric Research, Boulder, CO, USA

SnowMicroPen (SMP) Estimates of Snow Density on Sea Ice for Altimetry Applications

JOSHUA KING[1] AND STEPHEN HOWELL[1]

ABSTRACT

Observations of Arctic sea ice thickness are critical for evaluating the biological and climatological response to a warming climate. Basin scale observations of sea ice thickness have been routinely made with radar and laser satellite altimeters by estimating freeboard height and assuming a hydrostatic equilibrium. Using this approach, errors in snow load strongly influence freeboard height, ultimately impacting accuracy of the estimated ice thickness. In the in case of radar, spatiotemporal variations in the snow density influence velocity of the propagating wave, introducing further uncertainty in the freeboard estimates. Given the limited availability of *in situ* measurements, snow depth and density on sea ice are often assumed from climatology (i.e. Warren *et al.*, 1999) across large spatial and temporal domains. Here, we introduce an extensive suite of *in situ* snow property measurements on first and multiyear sea ice to establish a reference dataset and quantify potential errors in altimetry-based retrievals. Measurements of the SnowMicroPen (SMP), a high-resolution penetrometer, are used to retrieve high vertical resolution (5 mm) snow density profiles following the empirical model of Proksch *et al.*, (2015). Calibrated against a distributed set of snow pit measurements, the SMP profiles are used to describe spatial variability at horizontal scales of up to 100 m. Probability density functions generated from the SMP analysis allow development of radar and lidar altimeter error budgets. The results of this study will improve treatment of snow volume in current CryoSat-2 based-estimates of sea ice thickness as well as future altimetry missions such as IceSat-2.

REFERENCES

Proksch M, Löwe H, Schneebeli M. 2015. Density, specific surface area, and correlation length of snow measured by high-resolution penetrometry. *Journal of Geophysical Research: Earth Surface*, **120**(2): 346-362.

Warren SG, Rigor IG, Untersteiner N, Radionov VF, Bryazgin NN, Aleksandrov YI, Colony R. 1999. Snow depth on Arctic sea ice. *Journal of Climate*, **12**(6): 1814-1829.

[1] Climate Research Division, Environment and Climate Change Canada, Toronto, ON, Canada

Session #7: Remote Sensing Applications

Middle East Snow Cover Variability and Associated Atmospheric and Hydrologic Conditions

DAVID A. ROBINSON[1] AND M. NEIL WARD[1]

ABSTRACT

The Middle East is a region historically sensitive to climate variability and change and contains snowpacks that are important inputs to key regional water resources, including the Tigris-Euphrates river system. This presentation will examine the annual, interannual and decadal variability of the region's snowpack, and explore relationships between snow and associated atmospheric and hydrologic conditions. The presentation draws on satellite-based products, station data, and model reanalyses. Variation is summarized using space-time statistical techniques, as well as simpler regional indices, including Northwestern Iran / Southern Caucasus (NWIC, includes Zagros Mountains) and Eastern Turkey (ETKY, includes Taurus Mountains). The NOAA Interactive Multisensor Snow and Ice Mapping System tracks daily snow cover extent at 24 km resolution for 1999-present (primarily from visible satellite imagery). These data show that for both NWIC and ETKY, the mean snow extent peaks in late January with substantial coverage (~300,000 km^2 in each region), contracting to near zero by late June. A very large mid-winter interannual variance is also shown, implying substantial variation in hydrologic impacts during spring melt. Variability and decadal trends are compared with station snow depth reports (Global Historical Climatology Network – Daily). Strong agreement gives confidence in data quality, as well as, indicating high covariation of depth and extent. Connections with atmospheric variations and hydrologic impacts are recognized using reanalysis products and will be discussed as part of this presentation.

[1] Rutgers University, Piscataway, NJ, USA

Snow Estimation Capabilities for Military and Civil Works Applications and Operations

ELIAS J. DEEB[1], CARRIE M. VUYOVICH[1], JOHN B. EYLANDER[1], CHRISTOPHER A. HIEMSTRA[1], ANNA M. WAGNER[1], BLAINE F. MORRISS[1], TIMOTHY B. BALDWIN[1], AND JOHN J. GAGNON[1]

ABSTRACT

The U.S. Department of Defense (DoD) requires accurate terrestrial, atmospheric and environmental awareness on a global scale to support its national security, international development, and humanitarian functions. Specific snow information supports the Corps civil works mission, DoD's warfighting, and National intelligence functions. Snow is a spatially and temporally evolving medium that has a diverse set of impacts on DoD operations. While operational organizations provide general snow information with a global perspective, the time and length scales do not match tactical needs. Fine-scale spatial representation of snow requires observations or simulation of several snow characteristics including snow depth, density, albedo, stratigraphy, microstructure and temperature. The Remote Snow Assessment Team at the Cold Regions Research and Engineering Laboratory (CRREL) has addressed user needs through the use of a combined multi-sensor, modeling framework to improve global snow characterization and enable assimilation of remotely sensed observations. Over the past decade, CRREL has responded to requests for snow estimation capabilities, developed remote snow assessments for reach-back operational support, transitioned these capabilities to the Air Force, and investigated snow properties from microstructure to watershed scales through remote sensing, modeling, and data assimilation. We will present case study examples where the team has responded and answered questions regarding snow and its impacts on both military and civil works applications.

[1] U.S. Army Engineer Research and Development Center, Cold Regions Research and Engineering Laboratory, Hanover, NH, USA

Improvement of Airborne Gamma Radiation Snow Water Equivalent Estimations with Spaceborne Soil Moisture Observations

EUNSANG CHO[1], JENNIFER M. JACOBS[1], SAMUEL TUTTLE[2], RONNY SCHROEDER[1], AND CARRIE OLHEISER[3]

ABSTRACT

In the Northern Great Plains, accurate snow water equivalent (SWE) observations are needed to improve snowmelt flood forecasting. Since the 1980s, the National Oceanic and Atmospheric Administration's (NOAA) Office of Water Prediction's (OWP) airborne gamma radiation snow survey program has provided SWE observations to regional National Weather Service (NWS) River Forecast Centers (RFCs). Because the gamma radiation counts include the effect of both SWE and soil moisture (SM), the program requires a baseline soil moisture estimate in order to estimate SWE from their network of flight lines. Flight lines, which have spatial footprint sizes of 5–7 km^2, are typically flown once over bare soil in fall to obtain SM, and then flown over snow-covered ground in the winter. The difference between the fall and winter gamma radiation observations are used to calculate SWE. This approach assumes that soil moisture remains constant following the fall survey. However, rainfall events or drying after the fall survey, as well as drainage from snow freeze/thaw cycles during the winter, can alter SM. This study seeks to improve airborne gamma SWE measurements by updating the gamma fall SM with satellite passive microwave SM observations and modeled SM estimates immediately prior to winter onset. The passive microwave SM data used in this study are daily observations from the Soil Moisture Active Passive (SMAP) mission, available from 2015 to present. Results include relationships between gamma, SMAP, and model SM values, independent comparisons of the updated gamma soil moisture estimates, and improvements to gamma SWE relative to independent SWE measurements.

[1] University of New Hampshire, Durham, NH, USA
[2] Mount Holyoke College, South Hadley, MA, USA
[3] National Weather Service Office of Water Prediction – Minnesota, Chanhassen, MN, USA

Understanding Winter Temperature and Snowfall in the Anomalous Southern Appalachian Mountains: A 2017-2018 Winter Review

MONTANA A. ECK[1] AND L. BAKER PERRY[1]

ABSTRACT

Winter 2017-2018 will be remembered in the southern Appalachian Mountains for extreme volatility in both temperature and snowfall. The combination of frequent gulf low systems and multiple disruptions to the polar vortex allowed for an active snow season. An abnormally early start to the snow season and the longevity of cold air outbreaks across the region in December and January were directly at odds with the established positive North Atlantic Oscillation (NAO) and persistent La Niña, which are typically associated with positive temperature anomalies and less snowfall in the region. Winter came to a brief pause in February, with many locations breaking daily and monthly mean maximum temperature records. The warmth and infrequency of snowfall events in February directly affected local economies, with ski slopes having to close much earlier than anticipated. However, an unprecedented March brought a reinvigorated winter pattern with mean temperatures averaging several degrees colder than what was observed in February. Despite the early season chill, mean winter temperatures in the region finished slightly above normal, with the highest elevations receiving less snowfall than average. This winter season provides an opportunity to investigate how weather patterns in this region can rebuke the expected conditions during certain teleconnection phases, primarily the El Niño-Southern Oscillation (ENSO) and NAO. This work also highlights the importance of conducting research in the southeastern U.S. "warming hole" which has experienced an anomalous decline in temperatures since the 20th century, countering the warming seen globally.

[1] Appalachian State University, Boone, NC, USA

Winter Precipitation Forecasting at the Weather Prediction Center

DAN PETERSEN[1], GREG CARBIN[1], BRUCE VEENHUIS[1], MARK KLEIN[1], AND MIKE BODNAR[1]

ABSTRACT

The Weather Prediction Center works together with the forecast offices around the 48 contiguous states in the United States to produce both deterministic and probabilistic snow and freezing rain forecasts. The snowfall and freezing rain probability and percentile forecasts are posted online at *http://www.wpc.ncep.noaa.gov/#page=wwx*.

The forecasts employ the existing suite of operational forecast models and ensemble forecast members including the GFS, Global Ensemble Forecasts, NAM, Short Range Ensemble Forecasts, the ECMWF and the ECMWF ensemble members, and the members of the High Resolution Ensemble Forecast System.

These model forecasts are combined with human-produced snow and freezing rain forecasts for the next three days to provide a range of possible snow and freezing rain accumulations so people can assess how much snow and freezing rain will occur and when. The forecast graphics show whether the forecasts are in good agreement or vary widely.

Each day, the forecasts are collaborated with forecast offices around the country, who publish the amount of forecast snow online at *https://digital.weather.gov/*.

The offices are experimenting with displaying the snow probabilities with labels indicating the 'expected snowfall', 'high end amount', and 'low end amount'.

In the future, the Weather Prediction Center will coordinate issuance of Winter Storm Watches to notify the public of upcoming potential snow storms.

[1] Weather Prediction Center, National Weather Service, National Centers for Environmental Prediction

SNO-FOO 2018 – Alexandre Langlois

The 2017 SNO-FOO award is presented to Dr. Alexandre Langlois, Professor in the School of Applied Geoinformatics at Sherbrooke University. Alex wins the award while simultaneously serving as the 2nd Secretary, CA of the Eastern Snow Conference, a not-so-enviable feat!

Alexandre Langlois is a great collaborator, one you can always talk to and be frank with. Therefore, Alex always leaves his door wide open. He would never miss an opportunity to prove to you his openness. This time he was ready to show it again while spending time in the beautiful New Hampshire White Mountains during CrustEx. CrustEx took place in March 2018 and was mainly concerned with measuring and understanding the complex nature of New England snow. Especially snow that has a myriad of ice crusts from rain on snow events. In other words, it's where Alex loves to be scientifically even though he and Alexander Roy were several auto hours away from Sherbrook, Canada where both work.

One very cold morning during CrustEx, Alex's incredible openness caught up with him as he and Alexandre Roy were getting ready to leave camp early for radiometer measurements of freshly fallen snow. Little did both A's know what they were about to get themselves into. While backing out of a tight spot with their brand-new truck, a colleague flagged them down to check for room keys. Alexander Roy opened the passenger door of the truck to speak to the colleague while Alexander Langlois continued to drive backward into a snowbank pulling the door backwards and off the hinges like opening a can of sardines. I knew Alex wouldn't disappoint since he always leaves the door wide open! On the bright side, insurance covered the repair and they could close the door so both would not freeze to death while driving back to Canada that same day.

Presented by Ronny Schroeder

Past Recipients

The Eastern Snow Conference annual bestows upon a distinguished snow scientist who, in striving for excellence in snow research, contributes to an event of notable humor the highly coveted Sno-Foo Award.

Year	Winner	Affiliation
2018	Alexandre Langlois	Université de Sherbrooke
2017	Michael Durand	The Ohio State University
2016	Roger De Roo	University of Michigan
2015	Kevin Côté	Université de Sherbrooke
2014	Dorothy Hall	NASA GSFC
2013	Benoit Montpetit	Université de Sherbrooke
2012	Don Pierson	NYC DEP
2011	Ken Rancourt	Mt. Washington Observatory
2010	Kyung-Kuk (Kevin) Kang	University of Waterloo
2009	Rob Hellström	Bridgewater State University
2008	Steven Fassnacht	Colorado State University
2007	The Group of 9[1]	
2006	Andrew Klein	Texas A&M University
2005	Claude Duguay	University of Alaska-Fairbanks
2004	Chris Derksen	Meteorological Service of Canada
2003	Miles Ecclestone	Trent University
2002	Danny Marks	U.S.D.A.
2001	Brenda Toth	University of Saskatchewan
2000	Mauri Pelto	Nichols College
1999	Ross Brown	Meteorological Service of Canada
1998	Mary Albert	CRREL
1997	Jean Stein	Université de Laval
1996	Colin Taylor	Trent University
1995	Mike Demuth	NHRI
1994	Bert Davis	CRREL
1993	John Pomeroy	NHRI
1992	Tom Niziol	National Weather Service
1991	Terry Prowse	NHRI
1990	Kersi Davar	University of New Brunswick
1989	Gerry Jones	INRS-EAU
1988	Robert Sykes	SUNY
1987	John Metcalfe	Meteorological Service of Canada
1986	Peter Adams	Trent University
1985	Don Wiesnet	National Weather Service
1984	Barry Goodison	Meteorological Service of Canada

[1] Jimmy MacDonald (University of Saskatchewan), Bill Floyd (University of British Columbia), Chris DeBeer (University of Saskatchewan), Wendell Koning (Government of Alberta), Jaime Hood (University of Calgary), Dankia Muir (University of Calgary), John Jackson (University of Calgary), Sarah Forte (University of Calgary), Masaki Hayashi (University of Calgary)

Eastern Snow Conference
C/O Dr. Krystopher Chutko
117 Science Pl-Dept. Geography
Saskatoon, Sk, Canada S7N 5C8

ISBN 978-1-7138-0355-3

9 781713 803553

CURRAN ASSOCIATES INC.
proceedings.com

1st Symposium on Incoherent Light Sources

Loughborough, United Kingdom
8-11 April 1975